DARWIN IN SCOTLAND

DARWIN IN SCOTLAND

Edinburgh, Evolution and Enlightenment

J F Derry

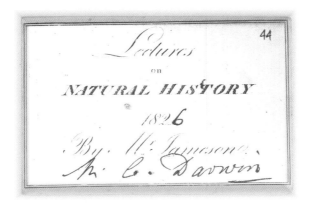

Whittles Publishing

Published by
Whittles Publishing,
Dunbeath,
Caithness, KW6 6EY,
Scotland, UK

www.whittlespublishing.com

Facsimile images (between pages 107 and 108) © 2010 Cambridge University Library
Edinburgh Reading List: transcription, footnotes and bibliography © 2010 The Darwin
Correspondence Project

Special thanks to Mrs Rosemary Clarkson for her assistance and to the Editors of the Darwin
Correspondence Project for access to unpublished material.

Photograph copyrights as accredited on individual images.

ISBN 978-1904445-57-9

Printed by Bell & Bain Ltd., Glasgow

To H, K & N

Charlie Is My Darwin

—attributed to Torn Rubbers (anagram of Robert Burns)

(sung to the tune of 'Charlie, He's My Darling' by Robert Burns)

'Twas in the year of 'fifty-nine
That Charlie told it clear:
'Tis nought but chance selection, boys,
The reason we are here, oh

Chorus:
Charlie is my Darwin, my Darwin, my Darwin,
Charlie is my Darwin,
Our one great pioneer.

He sailed the oceans of the Earth
To countries far and near,
He spent his youth in search of truth
A-followin' his idea, oh

(*Charlie is my Darwin, etc.*)

Although he'd borne the church's scorn,
He faced 'em without fear.
He spent his wealth and risked his health
To share his Big Idea, oh

(*Charlie is my Darwin, etc.*)

And poor old Bishop Wilberforce
So foolish did appear:
With Huxley's guile and Hooker's smile
They raised a mighty cheer, oh

(*Charlie is my Darwin, etc.*)

There's poor Lamarck, way off the mark,
And Wallace went astray,
Then Mendel's beans and talk of genes
Saw Charlie win the day, oh

(*Charlie is my Darwin, etc.*)

There's Dawkins, Gould and Lewontin
Still arguing the toss,
But none of them would yet deny
That Charlie's still the boss, oh

(*Charlie is my Darwin, etc.*)

Contents

Foreword by AC Grayling

Without question Charles Darwin's two years at Edinburgh University, passed there between the ages of sixteen and eighteen, were profoundly important for his later career. He was there ostensibly to study medicine, his father wishing him to follow in the paternal profession, but he so disliked the course that he paid it scarcely any attention, instead pursuing the passion for natural history that had been his primary avocation throughout boyhood at home in Shropshire.

Understandably, Edinburgh in particular and Scotland in general are eager to demonstrate their contribution to the development of Darwin's world-changing ideas. Edinburgh was then the 'Athens of the North,' the capital of the Scottish Enlightenment, home to a roll-call of genius that included David Hume, Adam Smith and Walter Scott. The grip of dour Calvinism had been sufficiently loosened for the fruits of intellectual liberty to appear, so that when in his second year Darwin met the anatomist Robert Grant, sixteen years older than himself and equally passionate about natural history, it was to encounter a freethinker who had been convinced by reading Erasmus Darwin, Lamarck and Étienne Geoffroy Saint-Hilaire, and by discussion with his geologist colleague Robert Jameson, of the truth of biological evolution by descent from common ancestors.

By Darwin's own account Grant's radical convictions on this subject had no more effect on him than those of his grandfather Erasmus, whose *Zoonomia* he had already read. 'I listened [to Grant on evolution] in silent astonishment,' Darwin later wrote, 'and as far as I can judge, without any effect on my mind. I had previously read the Zoonomia of my grandfather in which similar views are maintained, but without producing any effect on me.' Obviously, though, the convictions of both Erasmus Darwin and Robert Grant – and others such as Robert Jameson and the radical espousers of the 'materialism' which was said to be 'too common among medical students' – indeed had an effect on Darwin; those ideas had merely to wait upon the accumulated evidence of his *Beagle* voyage and his reflection upon it, to convince him in their turn.

Darwin's brilliant powers of observation were already evident in the collecting, exploring and bird-watching of his childhood at The Mount in

Shrewsbury. But they were fostered by his friendship with Grant, which was only occasionally clouded by Grant's habit of claiming credit for some of the discoveries Darwin made in their explorations along the shores of the Firth of Forth, studying marine invertebrates. In addition to Grant's tutoring Darwin benefited from Jameson's geological lectures, lessons in taxidermy with the freed slave John Edmonstone, meetings of the Wernerian and Plinian natural history societies, and the specimens provided by his favourite recreations of shooting and fishing; all this fertilised the soil from which his later work was to flower. When he went to Cambridge immediately afterwards he began collecting beetles with his natural-history-mad cousin William Darwin Fox, more influenced by this than the fact that his rooms at Christ's College were those that had formerly belonged to William Paley.

Darwin's Edinburgh years gave not just his interest in the study of nature, but his skill in carrying it out, a tremendous boost. The attitudes and beliefs of those he encountered there likewise had an effect, though not immediately. Had he gone straight to Cambridge he would still have been a naturalist; a clergy career for younger sons, or generally for youths with no inclination for medicine or law, was the most amenable for anyone in love with natural history – Gilbert White of Selborne is the type of the clergyman more in love with insects and birds than preaching. But Darwin learned more, and more quickly, at Edinburgh by his exposure to like-minded people, and to the wonderful opportunity of exploring the shores of the firth with Grant, than he would have done at Cambridge. And it has to be said, despite the hopelessness of proposing historical counterfactuals, that it is likely that modern evolutionary biology would be citing the name of Alfred Russel Wallace far more than that of Charles Darwin if Edinburgh had not been on his life's itinerary.

There is a general lesson in this. It is that climates of opinion, generally tendencies in the drift of ideas, opportunities for encounter between like-minded people, independence, the unfettered fostering of blue-sky interests, and the match of all or some of these things with a budding passion, make a powerful mixture. Of course original genius sometimes breaks through the antipathetic membrane of opposition and imposed orthodoxies. But Darwin's progress is like that of a ship given a good following wind and favourable tides; and Edinburgh is very much part of what wafted him on over the bar and out into the ocean of discovery.

AC Grayling
Professor of Philosophy,
Birkbeck College, University of London

Foreword by Stuart Monro

Science has always advanced through controversy and here, in *Darwin in Scotland*, stories of conflict and controversy are interwoven, separated sometimes in time but united in the 'hotbed of genius' that was Edinburgh in the years subsequent to the Scottish Enlightenment. As one of that group of scientists called geologists, I can testify to the fact that the natural world is wholly interconnected. The fundamental exploration of the nature of the world in which we live, how the earth works, cannot be looked at through the eyes of an individual scientific discipline, but rather through the connections between what we now call chemistry, physics and biology. The story of *Darwin in Scotland* exemplifies that interconnectivity and defines the legacy of Charles Darwin as not belonging solely to the biologists but shared with all the other science disciplines, yes, including geology!

The story brings together many of the pioneering names in science who influenced the thinking of Darwin but not all would be called biologists in today's terminology. James Hutton, the father of modern geology, gave Darwin a concept of the enormity of geological time, 'deep time'. It was a concept that emerged from observation of the relationship of the differing rock successions at that world famous unconformity, Siccar Point in Berwickshire, and a critical examination of what those observations meant. He concluded that the boundary between the two successions had to represent a period of time during which mountains could be built and eroded away – a time interval which must be very much longer than the 6000 years worked out by Archbishop Ussher of Armagh. Darwin's time in Edinburgh exposed him to this idea, a time scale in which changes through natural selection could be achieved.

This concept was as revolutionary as Darwin's own theories of natural selection and caused similar controversy, as did Hutton's ideas that rocks could be formed by magmatic processes; that is that they were one time molten. The substantiation of these ideas emerged from the quality of the initial observations and the deductive reasoning that followed. Perhaps Scotland provided Darwin not just with a framework of time, but also with some of the observational skills that underpinned his theories.

Darwin in Scotland also highlights some of the unsung heroes of this period and their influence on the young Darwin. James Croll is one such. A self taught man who began as the janitor at the Andersonian Institute in Glasgow (now Strathclyde University), he too was concerned with the big ideas in science, though he may not have realised it then. He it was who first identified the Earth's elliptical orbit round the Sun and that the wobble of the Earth's axis could control the amount of solar radiation that reached parts of the Earth's crust controlling the timing of glaciations. This concept was further developed by Milutin Milankovitch as Milankovitch Cycles which take their place today alongside the topical issue of anthropogenic climate change. It is a further, small example of the interconnectivity of natural science which was very much a part of Darwin's scientific philosophy.

The exploration of Darwin in Scotland reveals how much the evolution of big scientific ideas relies on an inter- or cross-disciplinary approach and the value of the critical evaluation of sound observations. Darwin's *Beagle* adventure was stimulated by a need for accurate cartography and a better understanding of volcanic islands, yet out of it emerged a truly big idea in biological science. This concept, though exemplified in Darwin's time in Scotland, is as true today as it was 250 years ago. The science base in Scotland today develops by connecting scientific disciplines and the lessons in *Darwin in Scotland* are relevant to the 21st century. To conclude with the Huttonian maxim, truly 'the past is the key to the future'.

Stuart Monro
Scientific Director, Our Dynamic Earth

Preface

Darwinian issues always seem topical, but never more so than now. At the time of writing there is great excitement at the bicentenary of Charles Darwin's birth and the 150th anniversary of the publication of the *Origin of Species*, both in 2009. The centres for this celebration seem to be Downe (naturally), the Natural History Museum in London where many of his specimens reside, and Cambridge, his second university and long-term publishers of his works and correspondence. Without wanting to detract from the events and activities at those three English institutions, there are a few additions that ought to be made in order to reaffirm Scotland's role in Darwin's history. In terms of Darwin's legacy, Scotland was his foremost influence. I am not maverick in my claim: Janet Browne, the science historian and Darwin biographer, wrote: 'Biographers regularly go back to Darwin's Edinburgh years, convinced the seeds of all his later thinking lie there – and to a large degree they are right'.

Firstly, the celebrations are late; 1856 was a momentous year towards the publication of Darwin's ideas. In that year he shared his ideas with colleagues, among them his closest friend and ally, and a Scot, Joseph Hooker; the response was largely positive, giving Darwin an important confidence boost. Having found acceptance of his ideas, Darwin received the nudge he needed to publish them a short time later when Charles Lyell, another Scot, alerted him to similar findings in the work of Alfred Russel Wallace, himself born in Welsh Monmouthshire, but to an English mother and a father claiming direct descent from William Wallace. Thus, they were Scottish sources that inspired publication of the *Origin of Species*, and, as we will discover later, it was a Scottish source from which that famous book's very title likely came.

Secondly, and more important for Darwin's skills as a naturalist, it was time spent in Edinburgh, including along the nearby coastline, and in the Highlands of Scotland that informed and trained the young man in preparation for his legendary voyage aboard the *Beagle* and his subsequent career in science. Darwin would not have developed his theories if he had not attended the University of Edinburgh. His formal tuition there didn't amount to much, but through interaction with his tutors, peers and extracurricular groups, Darwin was exposed to an ethos of naturalistic philosophy rooted in the Scottish

Enlightenment, and by direct descent, the Ancient Greeks. If he had bypassed Scotland, going instead straight up to Cambridge, his education would have been theologically based and unlikely to have given him the perspective that led him to question the prevailing doctrine. Of course, in recounting Darwin's well-known history, the emphasis herein may at times seem unusual, mainly because I have wanted to focus primarily on the part Scotland has played, and continues to play, in evolutionary science; the real opportunity for 2009 is for Scotland, and particularly Edinburgh, to rightfully claim its part in taking in the young man, and providing the people and places that formed the great naturalist.

Hence, this book partly makes a contribution towards that goal by investigating the influence of Darwin on contemporary biology, particularly as it is pursued at the University of Edinburgh. However, there is a snag. During the heyday of Victorian science, it was still possible to be a scientific generalist, something that Darwin was rather good at. Another example was his contemporary, Robert Chambers. Chambers, as an Edinburgh publisher (notably of *The Songs of Robert Burns*), a poet himself, prolific journalist and public figure, took up a keen interest in geology, and as was his wont – and befitting for a gentleman of his day – pursued fascinating interests. Chambers' fascination was an all-encompassing, overarching, general evolutionary, or development, theory. No mean undertaking, for a hobby! Well, times have certainly changed. As our understanding developed, we pigeonholed our knowledge within categories and sub-categories. Today, we continue that tendency to reflect our progress in the discovery of detail with the fractionation of science. For example, there are not many university departments fusing to form institutes of General Studies. Today, being a generalist is not so easy, and yet Darwin's work is as far reaching as it ever was, and it continues to pervade every contemporary biological discipline. The other role for Darwinian evolution puts it at the heart of the science–religion debate, as a counterpoint to contemporary Creationism and Intelligent Design.

Towards the aims of this book, it is now necessary to traverse those disciplines to piece together the extent and impact of Darwin's work. With this in mind, personal accounts were collected from luminaries associated with Edinburgh, that will hopefully reveal the influence of Darwin on their own interests. These interviews are reproduced as *verbatim* as possible, to avoid the pitfalls of quoting out of context. Therefore, huge thanks go to all the contributors for agreeing to take part and the many additional people with whom I have had discussions, including my publisher Keith Whittles, who was very patient in waiting for the reality following the promise. This book was born out of a larger project to be published as *The Dissent of Man: Darwin's Influence on Modern Thought*, which would be a best next destination when you have finished here. You may also be interested in companion pieces which have recently appeared in the *Edinburgh Science Magazine* (*EUSci*), the Edinburgh Royal College of Surgeon's *Surgeon's News*, the newsletter of the *ESRC Genomics Network*, and in Elsevier's *Trends in Ecology and Evolution* (*TREE*) and *Endeavour*. Thank you to Francis and Alison McNaughton and Angela Gibson who helped with transcribing the many taped interviews, and H, K & N for putting up with me for far too long while I worked on this. The writing was also made possible by a MacBook, NeoOffice, Mozart's operas and a sofa. A glossary and appendix of terminology are provided. The chapter epigraphs are all Burns quotes, which I hope you will agree represents a pleasant, if tenuous, fusion of the two anniversary Roberts: 250 years of Rabbie Burns (1759–1796), famously 'Scotland's favourite son', and 200 years of Charles Robert Darwin (1809–1882), one of Edinburgh's greatest 'sons of the city'.

JF Derry

1

SCIENTIFIC NATURALISM

Edina! Scotia's darling seat!

Edinburgh is often thought to be 'the pride of Scotland' – but not everyone would agree. Aberdeen, Dundee, Inverness, Perth, Stirling and particularly Glasgow are hard done by according to that verdict, each having had equal cause for a share of Scottish prestige. For example, Edinburgh wasn't even the first Scottish university. It followed establishment of academic institutions at St Andrews (1413), Glasgow (1451) and Aberdeen (1495), becoming Scotland's fourth university in 1583, at a time when England had only two (Oxford *c.*1167 and Cambridge 1209). However – and core to this book – Edinburgh's academic pedigree took the fore for its cosmopolitan perspective and central role in the Scottish Enlightenment during the latter half of the 18th century.

The Scottish Enlightenment arose out of a combination of social factors. The national programme of public education cemented by the 1696 Education Act achieved world-leading levels of adult literacy, and new international links throughout the British Empire, following the unpopular 1707 Act of Union, proffered opportunities for commercial prosperity. In addition an enduring 'special relationship' with France, itself undergoing a revolution in thought, brought a celebration of intellectual critique and rationalism, and paved the way for many of the advances in philosophical thinking to follow. The tangible products of this period included several that would be vital components of Darwin's education in Scotland: debating societies, scientific journals and the explosive rise of the book industry.

The Scottish Enlightenment movement was riveted in moral philosophy, history and economics. While it can be traced to the University of Glasgow (Francis Hutcheson 1694–1746, Adam Smith 1723–1790, Thomas Reid 1710–1796, and John Millar 1735–1801) – and there were the likes of James Dunbar

(1742–1798) attending the Aberdeen Philosophical Society – many of the key figures were at the University of Edinburgh (William Robertson 1721–1793, Adam Ferguson 1723–1815, and Dugald Stewart 1753–1828), or outside academia but living in the city (Henry Home 1696–1782, Robert Wallace 1697–1771, James Steuart 1713–1780, and James Anderson 1739–1808).

Outstanding in this latter group was David Hume (1711–1776), exponent of modern scientific empiricism (founded in 'experience and observation'), upon which all scientific truth has rested ever since. Indeed, without Hume every gem of science would glisten with a significantly lesser light. So, the next time you are passing by his statue on the High Street (also known as the Royal Mile), or his graveside in Calton Hill cemetery, a moment of reflective tribute may be due. Also up on that smallest of the 'seven hills of Edinburgh' you cannot fail to see a Parthenon replica, the National Monument. This building is otherwise known as Edinburgh's Disgrace and Edinburgh's Folly, for its unfinished state through exhaustion of funds. It is only one of the reasons why the city earned its name 'Athens of the North'.

That Athenian parallel with the classical period extends to the influence that Scotland had on the modern world, just as Greece had once enlightened the ancient world. If those lines of association run parallel, then others arrange in series, as lineages of inheritance from those ancient philosophies. Hume, for one, was directly influenced by Pyrrho (c.365–c.275 BC), founder of Pyrrhonian scepticism (in contrast to Plato's Academic scepticism). He was also indirectly influenced by the accumulation of independent, rational thought that had occurred across the two millennia since the beginnings of philosophy itself, a line going back to Thales of Miletus (624–c.546 BC). Thales was the first to attempt explanation of natural phenomena in rational terms. His approach replaced supernatural interpretations with scientific method and experimental protocol. This rationalism defines a clear line from Ancient Greece, via the Scottish Enlightenment, straight to the Victorian scientific naturalism (or methodological naturalism) to which Charles Darwin was first exposed at the University of Edinburgh.

Today the University of Edinburgh contains three colleges (Humanities and Social Science, Medicine and Veterinary Medicine, and Science and Engineering), consisting of 21 schools (e.g. Arts, Law, Science and Engineering, etc.). These sprawl across seven local areas (e.g. Central, King's Buildings including the Darwin Building and Library, Bush Estate) but also extend nationwide and internationally (reaching as far as China), and in total comprise more than 2000 buildings. However, it started from far more humble beginnings.

The original endowment in 1558, by Bishop Robert Reid of St Magnus Cathedral in Kirkwall, Orkney, led to a few centrally located buildings that became known as 'the Tounis Colledge'. This was followed in 1582 by a Royal Charter granted by James VI, son of Mary, Queen of Scots, which released to the University land where the church of St Mary in the Fields, or Kirk o' Field, had stood until the Scottish Reformation of 1560. The construction of today's architecturally grand central buildings (now constituting Old College) did not follow until around two centuries later, being completed only a few years before Darwin's arrival.[1] In 1789 the great neoclassical architect Robert Adam (1728–1792) re-founded the university buildings, a collection of random old buildings as they were then, and started building a double-courtyard system. He began at the northwest corner of Old College and the grand entrance range on to South Bridge, and continued until funding dried up because of Britain's involvement in the French Wars (1793–1806). This interruption in development left only one corner of the college useable. After the wars new funding and a new architect, the young William Henry Playfair (1790–1857), were found. Playfair had to compromise Adam's grand scheme, but by about 1820 the first buildings (Chemistry and Natural History) were completed, followed shortly afterwards by the east side. The north side was filled in with classrooms by 1824, just before Darwin's arrival, but there remained a great gap on the south side, until the new library was finished in 1829.

Why would Darwin choose to attend Edinburgh? The quick answer is that he didn't – his father decided for him, and Edinburgh was chosen because it was a family tradition to go there to study medicine. In fact, anyone pursuing a career in medicine was wise to go to Edinburgh, as since the mid-18th century it had been the best medical school. Oxford and Cambridge could give you academic teaching, but only if you were an Anglican, and they had no hospitals. London had no shortage of hospitals, but they had no university at that stage. Edinburgh had both a hospital and a university, allowing the parallel teaching of both theory and practice not afforded by anywhere else at the time. And this is the simple reason why both Darwin's grandfather and father had received their medical training at the University of Edinburgh – grandfather Erasmus (1731–1802) in 1755 (aged 22) and father Robert (1766–1848) in 1786 (aged 20) – both forging family ties and friendships while they did. One Darwin never managed to leave: Darwin's uncle, another Charles Darwin (1758–1778), died from meningococcal meningitis, contracted during a post-mortem, while studying in Edinburgh. He was buried in the Duncan family vault in the Chapel of Ease at St Cuthbert's Church which is beyond the west end of Princes Street Gardens, on the corner of Lothian Road and King's Stables Road.

Hence, there was an ever-increasing inevitability that Darwin would end up in Edinburgh, given its prominence as a seat of medical learning and the Darwin family connections, together with his father's growing exasperation with his lackadaisical son. In 1825 his father chastised him thus: 'You care for nothing but shooting, dogs, and rat-catching and you will be a disgrace to yourself and all your family'. Darwin's father also took action to rein in his wayward son, by removing him from his school in Shrewsbury, two years prematurely, for having achieved only 'ordinary' grades. Darwin's exit was recorded in his journal with undertones of understandable apprehension and self-pity: '1825 June 17th. Left Shrewsbury school for ever.— 16 years old'.

Darwin's mother, Susannah (née Wedgwood, 1765–1817), had died when he was eight years old, so he had nowhere to turn for motherly sympathy. To make matters worse, his male relatives had set vertiginously high standards – for example, his grandfather and father had been elected to the Royal Society while still audaciously young (aged 30 and 22, respectively). This goal Darwin would eventually meet (also when aged 30), but at that time everything must have seemed pointless and unachievable. Amazingly this family tradition would be kept going by three out of six of his own sons: Francis (when 34), George (also when 34) and Horace (when 52).

Reined in, and coming to appreciate the lifelong financial security promised by his father's wealth (a benefit of having married into the Wedgwood pottery dynasty), Darwin obediently spent the summer of 1825 showing his own promise as a physician tending to around a dozen of his father's patients. But at summer's end, and in a repeat of history, he was forcibly sent to study medicine at Edinburgh, just as his father had been forced by his grandfather. By treading in the footsteps of grandfather and father, the intention was for 'Bobby' to shadow his brother 'Eras' (Erasmus Alvey, 1804–1881), five years his senior and embarking on his year of hospital study from his own medical course at Christ's College, Cambridge. As soon as he was old enough, Darwin would also be expected to sit the medical examinations.[2]

When he did leave home, he had to pay seven pounds for his own travel northwards, to that new and alien environment: 'October. Went with Erasmus to Edinburgh'. But perhaps the prospect was not that daunting: Darwin was used to living away from home as a boarder in Shrewsbury. Additionally, he was accompanying his beloved brother with whom he shared interests in chemistry and the outdoors. Young, yes. But from his schooling and adolescent letters one can see that he was also self-aware and rebellious – useful buffers in a foreign land, a land that was on increasingly good terms with his English kind, ceremoniously marked by a visit of King George IV in 1822 (which incidentally raised the kilt, banned by the Dress Act in 1746 and reinstated in 1782, to an emblem of national identity).

2

DARWIN'S SCOTTISH ENLIGHTENMENT

Amang the rocks an' streams

The oldest influence on Darwin in Scotland came from the very rocks he studied there; in geological terms, modern humans came much, much later. Since appearing we have been philosophising about natural processes for only a comparatively short time. Nonetheless, and from the very outset, there has been an impressive range of disparate ideas. For example, Thales' protégé Anaximander of Miletus (*c.*610–*c.*546 BC) was the first to suggest that life had started from slime, and man had evolved, albeit directly from fish. Aristotle's (384–322 BC) theory of evolution somehow didn't involve any significant changes, living forms being maintained from one generation to the next in perpetuity,[1] and yet somehow incorporated the spontaneous generation of shellfish.[2] Empedocles of Agrigentum (*c.*490–*c.*430 BC) first recognised progression in living forms, and also wrote about the struggle for existence and survival of the fittest, as also did Titus Lucretius Carus (*c.*99–*c.*55 BC).

While some of these ideas would have been included in the university coursework, it is unlikely that Darwin could have been called a 'student' of these ancient philosophers at this stage. Aristotle's ideas would feature throughout much of Darwin's later influential reading – Malthus, Lyell and Herschel. Living in London and beginning to have his own early ideas on transmutation, he even made a note to diligently '[r]ead Aristotle to see whether any of my views are ancient'. But, while still at Edinburgh, more immediate to Darwin were his newly appointed teachers and newly met peers, in addition to the influences he already knew from within his own family: Erasmus Darwin, his grandfather, was known to him as an evolutionist. However, it is Darwin whom we call the true 'father of evolution' so what is it about his work that is more important to us than any other, ever since those Ancient Greeks? 'Natural selection,' you say. But even natural selection had been proposed previously: for purposes of explaining adaptation by those early philosophers Empedocles and Lucretius, then much later by Pierre-Louis Moreau de Maupertuis (1698–1759), Denis Diderot (1713–1784), James Hutton (1726–1797) and Isidore Geoffroy Saint-Hilaire (1805–1861). Natural selection had even been implicated in speciation by William Charles Wells (1757–1817), Patrick Matthew (1790–1874) and famously by Alfred Russel Wallace (1823–1913). So why is 'Charles Darwin' a household name?

The reason is because of the *way* in which he did it – his method. Darwin's use of his own acute powers of observation in tandem with the stringent scientific empiricism passed down by Hume, and by then considered mandatory by methodological naturalists, meant that his discoveries were so robust that he was able to both comment broadly about science at large and also comment in great detail within disciplines. Not only that, but his work was so mechanistically well rooted in the fundament of self-replicating life that it has relevance to every biological discipline, including those which hadn't even been established in his day. While the popularity of his ideas was no doubt initially assisted by the air of discovery in the Victorian era, paved by the Age of Enlightenment and the Scottish Enlightenment, when probably for the first time some factions were prepared to accept alternative explanations for the historical accounts inherited from their forefathers, the persistent longevity of his ideas is beyond doubt. No one can claim that they haven't survived the test of time.

Darwin's view of the world, and an ability to measure it, were instigated in Scotland: the University of Edinburgh seems to have been a bountiful source of influences on the young 16-year-old and he lapped up its strangeness and charm:

[Edinburgh]

Sunday morning.

My dear Father
As I suppose Erasmus has given all the particulars of the journey I will say no more about it, except that alltogether it has cost me 7 pounds. We got into our lodgings yesterday evening, which are very comfortable & near the College. Our Landlady, by name Mrs. Mackay, is a nice clean old body, and exceedingly civil & attentive. She lives in '11 Lothian Street Edinburg' & only four flights of steps from the ground floor

which is very moderate to some other lodgings that we were nearly taking. The terms are 1£–6s. for two very nice & light bedrooms & a nice sitting room; by the way, light bedrooms are very scarce articles in Edinburg, since most of them are little holes in which there is neither air or light. We called on Dr. Hawley the first morning, whom I think we never should have found had it not been a good natured Dr. of Divinity who took us into his Library & showed us a map, & gave us how find him: Indeed all the Scotchmen are so civil and attentive, that it is enough to make an Englishman ashamed of himself.

I should think Dr. Butler or any other fat English divine would take two utter strangers into his library and show them the way! When at last we found the Doctor & having made all the proper speeches on both sides we all three set out and walked all about the town; which we admire excessively; indeed Bridge Street is the most extraordinary thing I ever saw, and when we first looked over the sides we could hardly believe our eyes, when, instead of a fine river we saw a stream of people.

We spend all our mornings in promenading about the town, which we know pretty well, and in the Evenings we go to the play to hear Miss Stephens, which is quite delightful. She is very popular here, being encored to such a degree that she can hardly get on with the play. On Monday we are going to Der Fr. (I do not know how to spell the rest of the word). Before we got into our lodgings we were staying at the Star Hotel in Princes St. where to my surprise I met with an old school fellow whom I like very much; he is just come back from a walking tour in Switzerland, and is now going to study for [his degree?].

The introductory lectures begin next Wednesday, and we were matriculated for them on Saturday: we pay 10s. & write our names in a book, & the ceremony is finished; but the Library is not free to us till we get a ticket from a Professor.

We have just been to church and heard a sermon of only 20 minutes. I expected from Sir Walter Scott's account, a soul-cutting discourse of 2 hours & a half.

I remain Yr. affectionate son | C. Darwin.

Consequently, concerns about his immaturity would have been quickly put to rest. His new-found freedom cultivated pomposity, common among new undergraduates; his brashness was communicated to his sister Caroline, as indignation if slighted: 'I am very much surprised that Papa should so forget himself as to call me, a Collegian in the University of Edinburgh, a boy'.

There were plenty of young men who could claim that title: 'In the academic year 1825–26 there were in the University 2013 students in Medicine, Arts and Law, of whom 902 were in Medicine, and 250 of these came from England. In the year 1926–27 there were 1905 students, of whom 858 were in Medicine, and 215 of these came from England', although the relative seasonal immigration into the city would have been approximately half what it is today.[3] However, only about 20% would be expected to graduate through the 'Edinburgh model', described as 'a two-tiered model of medical education. Any student, whatever his previous background or studies, could attend classes. But only the privileged few, whose families could support them through three years of study, and whose Latin could sustain them through a thesis and examination, could graduate'. Solvent undergraduates from affluent backgrounds usually have funds for recreation. In Edinburgh, this has historically involved the opportunity to experiment with recreational drugs, something in which the Darwins partook:

> *The Gaieties of Edinburgh are now just beginning [...] do you remember Lady Harriet talking about inhaling <Ni>tric Oxide? Johnson has actually done it, & describes the effects as the most intense pleasure he ever felt. We both mean to get tipsey in the Vacation.*

All told, Darwin's first year was mostly notable for his sharing the fourth floor digs at Mrs Mackay's of 11 Lothian Street with Erasmus, and being invited to socialise with family friends. At school he had been an avid reader, loving poetry, particularly that of Byron, and the historical plays of Shakespeare. At Edinburgh Darwin continued that passion, he and Erasmus checking out more books from the library than all the other students combined, although some he admitted were not strictly course material: 'I have been most shockingly idle, actually reading two novels at once'. But after only four months Erasmus left and Darwin was abandoned to suffer the 'intolerably dull' lectures that he could still recall vividly, 55 years later, for his autobiography, 'with the exception of those on chemistry by [Professor Thomas] Hope', whereas 'Dr. [Andrew] Duncan's lectures on Materia Medica at 8 o'clock on a winter's morning are something fearful to remember. Dr. [Alexander] Munro [III] made his lectures on human anatomy as dull, as he was himself, and the subject disgusted me'.

A better idea of his daily timetable comes from the first part of Caroline's letter:

> *Edinburgh.*
>
> *Jan. 6th. 1826*
>
> *My dear Caroline,*
>
> *Many thanks for your very entertaining letter, which was a great relief after hearing a long stupid lecture from Duncan on Materia Medica. But as you know nothing either of the Lecture or Lecturers, I will give you a short account of them. Dr. Duncan is so very learned that his wisdom has left no room for his sense, & he lectures, as I have already said, on the Materia Medica, which cannot be translated into any word expressive enough of its stupidity. These few last mornings, however, he has shown signs of improvement & I hope he will 'go on as well as can be expected'. His lectures begin at eight in the morning. Dr. Hope begins at ten o'clock, & I like both him & his lectures very much. (After which Erasmus goes to Mr. Lizars on Anatomy, who is a charming Lecturer) At 12, the Hospital, after which I attend Munro on Anatomy– I dislike him & his Lectures so much that I cannot speak with decency about them. He is so dirty in person & actions. Thrice a week we have what is called Clinical Lectures, which means lectures on the sick people in the Hospitals ˜ these I like very much. I said this account should be short, but I am afraid it has been too long like the Lectures themselves.*
>
> *...*
>
> *I remain your af– dear Caroline, | C. Darwin.*
>
> *Love to Papa & tell him I am going to write to him in a few days*

Darwin had been unfortunate in his timing of attending the medical school. For three generations the Monro family had held the chair of anatomy, but even before Darwin had arrived in Edinburgh their domination in teaching was weakening. Between 1824 and 1825 the Royal College of Surgeons of Edinburgh accepted Robert Knox's plans and appointed him as conservator for their new Museum of Comparative Anatomy. This flamboyant and entertaining lecturer ran the recently deceased John Barclay's private anatomy school in Edinburgh's

Surgeon's Square from 1826, attracting many of Monro's disaffected students, including Erasmus who took private tuition from John Lizars.

The cost and supply of cadavers for anatomy classes was a problem to the students; before the Anatomy Act of 1832 only condemned criminals could be dissected. Erasmus, writing from London in October 1826, reported a similar shortage, in a rather objectified manner:

> The dissection is going on languidly [...] there is but one subject come in yet & there are six engaged before the one I have put my name down to: they are cheap compared with Edinburgh being £8 8" [eight pounds and eight shillings] which however when it comes to be multiplied three or four times is a heavy draw back.

An unorthodox solution to the problem was instigated by two Ulstermen who came to Scotland to work as labourers on the Union Canal. The death of an old and indebted lodger led, via the graveyard, to Knox's door, for his reputation of paying more than Monro: because it was summer, and corpses were less fresh, they received £7. 10s. (seven pounds and ten shillings). From there on, Burke and Hare didn't wait for their victims to die of natural causes. Their 16 murders were carried out from the autumn of 1827 until their capture in November 1828. Their final victim earned them £10, Knox's winter rate. Hare turned King's Evidence and fled the country. Knox was vilified by the rioting masses – 'Burke's the butcher, Hare's the thief, Knox, the boy who buys the beef!' – but other than receiving mild threats and minor property damage he was exonerated, and remained successful for most of his remaining 33 years. In contrast, in January 1829, watched by 25 000 spectators, Burke was hanged in Edinburgh's Lawnmarket, and then, fittingly, dissected, by Monro, 'in the University of Edinburgh's medical school, and his skeleton still hangs there for students to observe. His skin was made into various items, including [a] pocket book. The pocket book is currently on display in the Surgeons' Hall Museum'.[5]

Only a matter of months before this particular couple of resurrectionists were at large, dealing in the dead, Darwin was having to attend unanaesthetised operations on the living. Two in particular he found more disgusting than Monro and his human anatomy. It was an experience he never repeated, nor needed to: 'The two cases fairly haunted me for many a long year'. These operations and general boredom probably figured in Darwin's intention to go home early before the end of semester. His journey was to include return via Glasgow – 'I mean to go as far as Glasgow by the canals and from thence on "terra firma" to Shrewsbury, but sending my books per sea' – a show of independence likely inspired by Erasmus' account of his trip there via canal, in early March, involving bottled beer and toddy. This suggestion was not well received. His father sent protestations, first indirectly via Darwin's sister Susan:

[Shrewsbury]

[27 March 1826]

Monday Evening

My dear Charley
I was very well pleased to receive your last letter as I hope you are
getting in better humour with Edinburgh now that Spring is come. Do
go some day to see Roslyn Castle, which I believe is within a few miles
from Edinburgh, & it is very well worth seeing. At present they have
made a Diorama of it, in London. My reason for writing so soon is,
that I have a message from Papa to give you, which I am afraid you
won't like; he desires me to say that he thinks your plan of picking &
chusing what lectures you like to attend, not at all a good one; and as
you cannot have enough information to know what may be of use to
you, it is quite necessary for you to bear with a good deal of stupid &
dry work: but if you do not discontinue your present indulgent way,
your course of study will be utterly useless. Papa was sorry to hear
that you thought of coming home before the course of Lectures were
finished, but hopes you will not do so.
...

I am so sleepy I must wish you Good night | ever yr affect | Susan

When Darwin proved unrelenting of his plans, Susan's tip-off was followed
up a few weeks later by a joint letter from other sisters Emily and Caroline,
conveying relief at his acceptance of their father's will, which had delayed his
departure, and had probably wanted to thwart it entirely:

Thank you dear Charles for your very kind affection<ate> letter, you
can not tell how much I value your love. I am very glad we shall have
you at home again soon. & Papa is very glad that you have remained
to attend all the Lectures, as he is sure they will be useful to you. He
wants to know if you are thinking of any excursion in your way home
to see any new country that you did not pass through as you went to
Edinburgh. Papa is rather surprised at your going to Glasgow as there
is nothing worth seeing in your way there, & he should have thought
it a better plan to leave as much of your heavy luggage as you can

Giant of the Scottish Enlightenment and a primary exponent of empiricism, David Hume's statue is now prominent in front of the High Court Building on the Royal Mile (High Street) of Edinburgh. His grave is marked by an impressive monument in the Old Calton burial ground.

Above left: A plaque marks Darwin's lodgings at 11, Lothian Street, the building now replaced by the National Museum of Scotland.

Above right: As an assistant at the University Museum Darwin had use of a desk, probably where Edmonstone taught him taxidermy, and likely located in the annex below this spiral staircase.

Below: The Old College of The University of Edinburgh, much unchanged since its original design by Robert Adam, and completion by William Playfair (with Adam's dome being added by Robert Rowand Anderson in 1877).

Photographs © Melinda T. Hough 2010

in Edinburgh as you will return there & so not have the plague of the carriage of it backwards & forwards.

To clear his head of anatomy and likely put some distance between himself and the long reach of home, Darwin took an energetic walking holiday in North Wales with two friends during the summer of 1826, and became further inspired to observe wildlife – and record his observations – from reading the Reverend Gilbert White's *The Natural History of Selborne*. After the summer he returned to a more sociable second year in Edinburgh. He became acquainted with surgeon and future geologist William Francis Ainsworth (1807–1896); William Macgillivray (1796–1852), the conservator of the Royal College of Surgeons Museum where he attended talks; Mr Leonard Horner (1785–1864), Charles Lyell's future father-in-law, who 'also took me once to a meeting of the Royal Society of Edinburgh, where I saw Sir Walter Scott in the chair as President'; and a freed black slave John Edmonstone, who fuelled his wanderlust while teaching him taxidermy: 'I am going to learn to stuff birds, from a blackamoor I believe an old servant of Dr. Duncan: it has the recommendation of cheapness, if it has nothing else, as he only charges one guinea, for an hour every day for two months'. This was a sound investment as it would prove invaluable on the *Beagle* voyage. Their racial differences would not have been an issue as Darwin had been raised strongly anti-slavery. A psychological analysis of his letter writing at the time has suggested that exposure to Edmonstone's life story may also have given him some perspective on his own position:

> On April 8 [1826], in a letter to his sister Caroline thanking her for her 'kind letter', Charles wrote two sentences that reveal a maturing, introspective perspective, and a self-deprecating stance that reappeared throughout his life: 'It makes me feel how very ungrateful I have been to you for all the kindness and trouble you took for me when I was a child. Indeed I often cannot help wondering at my own blind Ungratefulness'.[6]

Also notably, he joined the student-body Plinian Society, established by geologist Professor Robert Jameson (1774–1854) and dominated by radical naturalists. Prompted by future physician John Coldstream (1806–1863), Darwin presented a paper there, on the marine biology of the Firth of Forth which he had sampled from dredge boats at Newhaven, and in the tidal pools near Prestonpans. These seaside excursions were made with possibly Darwin's single most important influence, the anatomist Dr Robert Edmund Grant (1793–1874).

While 17 years older than Darwin, and therefore an unlikely companion, Grant was a graduate of Edinburgh's medical school, but had not continued on to practise as a physician. In preference, he was an avid naturalist, and so perhaps a promise of future prospects for the disillusioned medical student. Grant was also friends with the as-yet untarnished Robert Knox and the French naturalist Étienne Geoffroy Saint-Hilaire (1772–1844), forming a direct international link between the University of Edinburgh's school of anatomy and the hub of Lamarckism in Paris for which they shared a passion. With Lamarckism came an acceptance of anti-Aristotelian transmutationism, a radical and challenging position for anyone coming from outside progressive Scotland. However, thanks to his freethinking background, Darwin was already aware that alternative ideas existed in natural history, even if he hadn't readily grasped their evolutionary significance:

> I knew him well; he was dry and formal in manner, but with much enthusiasm beneath this outer crust. He one day, when we were walking together burst forth in high admiration of Lamarck and his views on evolution. I listened in silent astonishment, and as far as I can judge, without any effect on my mind. I had previously read the *Zoönomia* of my grandfather, in which similar views are maintained, but without producing any effect on me. Nevertheless it is probable that the hearing rather early in life such views maintained and praised may have favoured my upholding them under a different form in my *Origin of Species*.

3

MENDELIAN RATIOS

In every knot and thrum

Nikolaas 'Niko' Tinbergen (1907–1988) famously framed Darwin's natural selection, the central thesis of the *Origin of Species*, within his scheme of Four Questions [of Animal Behaviour], which were inspired by Aristotle's Four Causes.[1] Also seeking answers to the question 'Why?', the first of Tinbergen's 'ultimate explanations' dealt with function and tied together survival, reproduction and adaptation, mediated through natural selection. This grouping corresponds to Aristotle's Final Cause that teleologically seeks to explain purpose. Scientific reductionism then seeks to identify what quantity is being selected naturally: species, group, individual, cell or gene, because this would then reveal the objective purpose for life. For example, you will have heard the expression 'for the good of the species', but is it whole species that evolution is benefiting? At the opposite end of the spectrum, the ultimate purpose of the selfish gene, according to Tinbergen student Richard Dawkins, is maximal self-replication whereas, somewhere within the middle ground of genetic relatedness, William Hamilton's kin selection incorporates altruism towards benefiting reproduction of one's relatives.

The ongoing debate about this unit of selection was made possible by the modern evolutionary synthesis which integrated Mendelian genetics with Darwinian evolution. Before, it was difficult to talk in terms other than selection of individuals and groups 'for the good of the species'. But genetics introduced a smaller scale, within individuals and groups of similar individuals, right down to the level of the single gene. Later, we'll come to see how modern evolutionary biology and genetics have made it possible to peer deep within the individual, by using and developing Darwin's original ideas.

It's a fascinating thought what Darwin might have done had he been privy to information on genetics and especially gene flow. Alas, one could say it was 'over his head'. It's little known, but as an assistant at the University Museum he had use of a desk. Although John Edmonstone and Darwin lived in the same nearby street, this is probably where they came for their lessons on taxidermy, and also where Darwin's head was filled with tales of the Guyanan jungle. The desk sits at the bottom of a stairwell in a building at Old College that since 1975 has housed the Talbot Rice Gallery. Now, that staircase is special. It is an elegant, sweeping, spiral one. So, as he leant back, hands behind his head, looking absentmindedly cupola-wards, Darwin's gaze could not help but follow the parallel lines of a double helix towering above him. What sweet, delicious, wrought irony. Artist and author Alistair Gentry was the most recent incumbent of Darwin's desk-space:

> [Darwin] did formerly sit underneath the stairs when the Talbot Rice Gallery was a library/study hall, although I've not heard or read of any reason why he was there and not in the main room. Maybe he was already feeling like an outsider, or was already determined to make himself one. I was just there for lack of anywhere better to put me, and probably far more of a university outsider than Darwin ever was. I was always aware that the university would never have had me as a student in the first place [...] The whole of the old gallery, stairwell included, is a beautiful environment, but to me being in there every day it also became a kind of domestic place where I'd meet people and have a cup of tea. I also held events and workshops there related to my work, so it was often temporarily 'my' space [...] it was actually the place (and where I sat was specifically the physical location) where he grew to hate the university, totally lost interest in studying medicine and being a medical doctor, went against his family's wishes for his career, and started on the path that ultimately undermined the scientific and religious orthodoxy of the time [...] I definitely had a love/hate relationship with the university as an institution [...] everybody, throughout the university, seems to waste an inordinate and disgraceful proportion of their time talking about sheer bullshit, things that barely even matter in the context of academia or specialised research, let alone in the context of life as a whole. I was certainly very frustrated by the inertia, complacency and protocol of such an ancient, sprawling institution, and I could easily connect with anybody who'd spent a lot of time sitting (mostly) alone under the stairs with their work, their thoughts and the feeling of being full of ideas that the rules don't allow for. I doubt I'll come up with anything as revolutionary as evolutionary theory, though.

While he actively opted out of mainstream education at the university, it's unlikely that Darwin was already having ideas that would challenge the orthodoxy while still in Edinburgh. Therefore, the next few chapters will loosely follow the development of evolutionary knowledge in a roughly chronological order, as recounted predominantly by leading contemporary biologists at the University of Edinburgh, beginning with how it was available to Darwin before it became 'genetically modified'. To commence, Niko Tinbergen's other most famous student, renowned animal behaviourist and broadcaster Aubrey Manning, will share some of his insights:

Well, I first came to Darwin really when I started my DPhil at Oxford on the reaction of bees to flowers and what patterns on flowers they responded to, and I don't know whether it was Niko Tinbergen my supervisor who put me onto this; it could well have been J.B.S. Haldane and Helen Spurway, who had been very kind to me as a student at UCL [University College London] and kept in touch with me as I moved to Oxford. Anyway, somehow I began reading Darwin on plants. I read in particular the wonderful book *[On] the Various Contrivances by which [British and Foreign] Orchids are Fertilised by Insects*, and then one on *The Effects of Cross and Self Fertilisation in the Vegetable Kingdom*. That, I think, was probably the book I enjoyed the most.[2] Then also, and interestingly enough, I have never been able to re-find it, but if you look carefully into Darwin's observations, when he was growing pin- and thrum-eyed primroses in his greenhouse, he's found, what we now know, that there is a single gene involved: pin is the dominant and thrum the homozygous recessive. And somewhere in Darwin[3] he counts an F2 [second filial hybrid] generation [a cross between two first filial hybrids], and says that the progeny approximates in a proportion of 3-to-1. Now, he makes no further comment on this at all, he's simply stating a fact that there were three times as many pin-eyed as there were thrum-eyed in the F2 generation. I read that once and I remember being ... well it was as if somebody had punched me in the gut, it was just incredible! But of course he made nothing of it at all.

Did you know Mendelian type ratios from your education?

I knew them yes. I was a graduate, so I knew all the genetics, but he of course had all these ideas of pangenesis and so on. Mendel, I understand, sent reprints to Darwin because he greatly admired him. There's no evidence that Darwin ever read them and in particular of course they were in German and it was his wife Emma who read the German literature for him. That's fascinating, but relevant to me in particular were some observations that he recounts, I think in *The Effects of Cross and Self Fertilisation in the Vegetable Kingdom*, watching bees pollinating plants, visiting plants, and he discusses how bees clearly have a knowledge of the group of plants on which they are foraging, and head off not steering by stimuli directly from the plant, but from their memory. And Tinbergen had put me onto making further observations on this kind of phenomenon which I did, and I think Darwin was the first person to describe this. Also, I loved the way Darwin writes. I love Victorian novels anyway, and I love his prose style. To me the ending of the *Origin of Species* is a most lovely passage: 'It is interesting to contemplate an entangled bank...'.[4] Very, very beautiful. Funnily enough, when my good friend David Wood-Gush died he asked to have that passage read out at his funeral [...] and it immediately sparked off something in me because in it he says it is wonderful to consider all these multiple forms of life; he mentions the Creator, the Creator breathes life into one or many of them,[5] and I thought, 'that sounds funny to me', because I knew that Darwin, certainly quite early on, lost his Christianity really when his daughter Annie died.[6] So I made enquiries with dear old Arthur Cain who was then alive in

Liverpool, a great expert on Darwin, and sure enough those words were not in the first edition of the *Origin* [of Species], life breathed into one form or into many, the Creator's not mentioned. He put that into the sixth edition for Emma I think. She was quite religious and remained devout. I love that and I think one of the proudest moments of my life was when I was just a DPhil student writing up my first year report, and, well, I didn't know how to write, I had never written a paper before I suppose. Helen Spurway wrote me back, I tried to find the letter, I can't find it, she said: 'I gave your report to Haldane to read and he brought it back and said, "This chap writes like Darwin"!' And I thought, 'God, this is fantastic. I can't believe it's happening to me'. Needless to say I was cut down to size later on because when I submitted my thesis it was failed on the first attempt because the examiners did not like my style at all! So there it was. However, never mind, I got it the second time through.

Well, those are some of my fondest memories of reading Darwin. Of course, for animal behaviourists people now go to *The Expression of the Emotions in Man and Animals* and of course *The Descent of Man*, [and Selection] in Relation to Sex. I can't say I ever read those at the time when I was really building up my knowledge of animal behaviour. They have been profoundly influential of course. The thing on sexual selection where he had this great, I wouldn't say argument with Wallace, but they certainly differed with one another about whether the females are choosing males for aesthetic reasons or whether they are linked to fitness in some way or other – I can't quite remember how the arguments go to tell you the truth. I had to review a book by, what's the name of that bright woman, interesting woman at LSE [London School of Economics] who's written on evolutionary biology? Helena, Helena somebody or other, maybe Cronin; Helena Cronin I think it is. Anyway she wrote a very interesting book called *The Ant and the Peacock* which was about the, maybe you know it, yes, it's a very interesting book, a really scholarly account of the origins of sexual selection and also of altruism, the evolution of altruism. And I had to review that at length in some journal, and I remember going into the arguments there. Of course, it's very interesting that Darwin and Wallace had this 'mutual admiration society' for so long, but I really have no truck with this idea that Darwin was involved in some kind of shabby way of getting around being scooped by Wallace. I don't believe that at all.

It seems that he was too honourable in everything else he did.

He was far too honourable a man. He was a product of his generation. He was a gentleman. You're right. My father was a cockney, and the cockneys had a term, they referred to somebody as 'a Gent', and that meant something serious. That was not a flippant term you used, you didn't call someone a 'Gent' lightly. In fact of course, Wallace was a huge admirer of Darwin, and their correspondence shows how much they owed to one another. Darwin is prolix in his praise for Wallace's early observations, but at the end they finally parted company didn't they, over the origin of the human brain, where Darwin stuck to his guns, it's got to be natural selection. There is no other force shaping us, that and our environment. But Wallace felt that, strangely, the human brain must be God-endowed, because savages had brains like

English gentlemen, and they didn't need brains like English gentlemen. How could he be so blind! He was such a brilliant man. And he didn't see that these Malaysians who he had been living with, the people in South America, he couldn't see that they were human...

That is the irony, because Wallace spent far longer in the wilds than Darwin, and it is incredible what Darwin did in those few years…

Far longer, yes absolutely. After that he [Darwin] scarcely left Kent! He certainly went to Shropshire and Worcestershire, because he took Annie to get the waters at Malvern, and alas it didn't do her any good, since she most certainly had typhoid. Darwin, of course, he had great trouble, you remember his comments on the Tierra del Fuegans, 'it was hard to believe that they were the same species', but he knew that they were. He, I think, was prepared to see that although they lived in what to him looked like an utterly savage manner, they had their own subtleties which were not obvious.

Unlike Wallace, Darwin comes across as being very much a humanitarian, because he was so far ahead of his time and he applies himself with sincerity; he also comes across as being way ahead of Victorian attitudes.

But you see Wallace was a socialist. He was a socialist of the old school. He was a great admirer of Robert Owen and the New Lanark, these Utopian societies. He wrote books about social welfare. All of this, and yet somehow he believes that these 'savages' didn't require the same brain that we had, and yet he knew they did have the same brain as we have, and therefore, it couldn't have been natural selection. Bizarre!

Evolutionary psychology is a relatively new field which has allowed some quantification of the brain function, but it is still a pseudo-science in some people's opinion. I don't know if Darwinian science is becoming more specialised, or less specialised.

Well it's certainly possible to have a synthesis, I agree. It is possible to bring them all together; I mean evolutionary biology and behaviour have come together and there have been marvellous insights gained from the old sociobiology. Evolutionary psychology sometimes goes way over the top, and when psychology tries to be desperately scientific and thinks that counting things is what you need to do, it's missing the whole point. Human beings function at many different levels, and one of the levels we function at is the level of our minds and our cognition, our emotions and so on, so we need to study them at that level, and that may mean you can't count everything. You can't count how much anxiety I feel on a scale of 1 to 10. So I think that psychology has gone down some pretty crazy paths, and I find that some of the modern evolutionary psychology, it disturbs me in a way, because I think it takes far too little account of cultural effects. On the other hand, I have to agree that sometimes our reactions are extremely simplistic, and at an unconscious level we may be operating in a much more straightforward fashion. I'm fascinated by this, needless to say that you hear of these studies, you see them illustrated, and you don't really get to read them and follow up, but I remember a man, he was Dutch I think, he was taking photographs of young women in a disco in Amsterdam.[7] He would say he and his assistant were running a study on physical type and blood sugar or something like that. This was a phoney. What they would do, they would photograph these women straight out of the disco and he would ask them to give a sample of saliva to measure their blood sugar; actually he was measuring their hormones of course, so he could estimate where they were in the menstrual cycle. And the story showed that the closer they were to ovulation, the more flesh was visible. Now do I believe this? Do I believe this? I mean, I don't want to believe this.

Do you also read other voyages like Humboldt?

I have read Wallace. *Malay Archipelago* is just a wonderful book; he writes beautifully. It is an absolute pleasure to read it. In his correspondence it's wonderful because he switches from the absolute detail, 'I think you are incorrect in specifying these details of the penis of this species of barnacle which I have found to lack spines on that particular segment', and so on [...] to talking in generalities about the effects of natural selection, [...] how it might be that acquired behaviour might become incorporated: he's got this thing about Robinson Crusoe's parrot, a lovely idea. He had this idea that if Robinson Crusoe teaches his parrot to say 'Good morning, Sir' and the parrot breeds, and he keeps breeding from the ones that say 'Good morning, Sir' quickest, [...] eventually Robinson Crusoe's great, great grandson has got this 20th generation of parrots and they're saying 'Good morning, Sir' when they first hatch. I mean he speculates about it in a quite delightful way.

But also I find it touching to read about Darwin as the family man, and about the tragedy of his little daughter Annie. I was in Malvern recently and I [went] into the priory churchyard and found her grave, and it simply said on it, 'Anne Elizabeth Darwin, 1841–1851. A dear and good child'. Nothing more. Emma wanted of course 'In God's bosom', but not for Darwin. I was also pleased to see that there were fresh flowers on it.

The letters he writes about that time are really touching because he's trying to keep up his scientific correspondence, and he's saying, 'You must forgive me for the great delay in responding to your interesting letter about these plants, but we have suffered a great loss'. His correspondence is staggering, I think it's going to be more than 20 volumes. It's a major undertaking. I don't know how Darwin managed his correspondence. Did he have a secretary who wrote them? How did we get copies of these letters? I mean, was there carbon paper? You couldn't go up and take a photocopy of it. How did they do it? I don't know.

You obviously know Down House and you've seen the documents there.

I've never been there! I must go! But it's a place of pilgrimage for me and I intend to go down to Canterbury ... and walk on the sand path. Now, a friend of mine, Richard Andrew, a behaviourist from Sussex, went there late one afternoon, with his wife, and the guy there, the warden, saw how captivated Richard was, and he let Richard sit in Darwin's chair, and Richard said his scalp crawled!

4

MODERN EVOLUTIONARY SYNTHESIS

Lord grant that thou may ay inherit

Darwin worked on the *Origin of Species* mostly while based at Downe.[1] The
first version took 17 years to produce. It could have taken much longer
had Darwin not been forced to publish with the London-based firm
of Edinburgh-born John Murray; he had intended it to be part of a bigger 'Big
Book' on species, which he had provisionally called, simply, *Natural Selection*.
Before Darwin's shortened account of natural selection did make it into print,
the widely accepted evolutionary mechanism had been one of the heritability
of organs based on 'use and disuse': like a bodybuilder's biceps, utility leads to
development and enlargement, and disuse ends in necrotic withering.

This within-a-single-generation 'inheritance of acquired characters' became
popularised by Jean-Baptiste Pierre Antoine de Monet, Chevalier de Lamarck
(1744–1829), most comprehensively in a section on 'L'influence des circonstances'
(The adaptive force) of his *Philosophie zoologique* – hence 'Lamarckian
inheritance'. Yet, like many people who have become inextricably linked to
their source of fame (Darwin didn't single-handedly discover evolution, Miles
Davis didn't single-handedly invent jazz-fusion[2]), Lamarck's inheritance was
adopted and adapted from the mainstream view of inheritance. He especially
benefited from the encyclopaedia of polymath and fellow Frenchman Georges-
Louis Leclerc, Comte de Buffon (1707–1788), who managed to publish 35 out
of the 50 planned volumes of his great *Histoire naturelle, générale et particulière*.

Lamarck presented those commonly held ideas on inheritance comprehensively and attractively to a wider audience, hence his legacy, although the
whole package (comprising the adaptive force and the complexifying force,
where adaptation drives evolution from the simple to the complex) would be
better called Lamarckism. This school of transmutationism is what Grant

subscribed to, and was written about in the first volume of Robert Jameson's *Edinburgh New Philosophical Journal*,[3] readily available to Darwin when it came out in 1826.

Lamarckian inheritance also incorporates aspects of learning: the educated offspring of Robinson Crusoe's parrot, or, similarly, the drooling progeny of Pavlov's dogs. It would not however include learning capacity *per se* or instinct, which would come under ontogenic evolution, also named the 'Baldwin effect' after psychologist and philosopher James Mark Baldwin's (1861–1934) paper 'A new factor in evolution'. However, as Lamarck's second volume of his *Histoire naturelle des animaux sans vertèbres* was being published three-quarters of a century earlier in 1822, the path to understanding how inheritance, and thus evolution, works was being paved, not a million miles away.

Johann Mendel (1822–1884) was born on 20th July to Anton and Rosine, a farming family living in Heinzendorf which was then part of Austrian Silesia. Unbeknownst to him, his gardening boyhood was perfect preparation for his future life as a monk. Initially he had entered the Augustinian monastery in Brünn (Brno) in 1843 as a novice, largely to escape the poverty of his family life. Then, ordained in 1847 and taking the name 'Gregor', he spent a frustrating decade trying to qualify as a teacher, especially of mathematics and biology, but was hampered by his nerves during examinations. Lacking direction, he began cultivating pea plants, eventually about 28 000 of them, noticing varietal differences in seed shape and colour, stem length and plant height.

Gregor Mendel published his results as 'Versuche über Pflanzen-Hybriden' in 1865, but the paper went unnoticed (until the early 1900s when his work was independently verified thrice, revealing the correct accreditation more than 20 years after his death). It is commonly thought that he likely sent a copy to Darwin; the impact to his work is an intriguing prospect. It was even rumoured that an unread copy survived at Down House, which would have been a travesty, but none, nor record of any, has ever been found amongst Darwin's library (Nino Strachey, former Curator, Down House, cited in Sclater 2006). Sclater does however also identify two publications that were in Darwin's possession that could have provided alternative avenues to Mendel's work: *Untersuchungen zur Bestimmung des Wertes von Species und Varietät* by Heinrich Karl Hermann Hoffmann (1819–1891) and *Die Pflanzen-mischlinge* by Wilhelm Olbers Focke (1834–1922). However, the lack of detail, the emphasis of the material in these books, and their timing – coupled with Darwin's omission of comparison with his own, similar findings – suggest that their information was parsimonious or uninspiring. In any case, future opportunity would have been short-lived: in 1868 Mendel was promoted to monastery abbot, essentially ending his research when he 'grew too old and fat to bend over and cultivate his pea plants'.[4]

It is part fascination, part frustration, that makes one ponder the consequences of Darwin having been privy to Mendel's results. As we have seen, he would have immediately recognised the ratios from his own trials, and yet it is clear that no such meeting ever took place. Nonetheless, there is a wonderfully tantalising entry in Emma Darwin's diary[5] for Monday, 26 November 1866 which reads: 'Mendel. ottett !! Wilhelmj Pop. concert'. Does Emma's diary, full of shorthand and codes, suggest that they planned to meet? Alas not. That evening they were simply to enjoy a concert of Mendelssohn's Octet in E-flat major, op. 20, about which Emma was clearly excited, using double exclamation marks. This series of 'Monday Popular Concerts' was particularly auspicious. Advertised in *The Times* and attended by Prince Leopold, it was to be held at London's principal concert venue, St James' Hall. It was also the first English tour for the 21-year-old violin soloist August Daniel Ferdinand Victor Wilhelmj (1845–1908), a prodigy hailed as 'the German Paganini'. The concert was also to include Beethoven's Romance in F, and Mendelssohn's C minor Trio performed with Charles Hallé (pianoforte) and Piatti (violoncello).[6] The tour was a sensation. The only criticism from the press, levied by *The Morning Post*, suggested an alternative spelling for the violinist: 'Wilhelmi would be more simple and intelligible to English eyes'.

Such a meeting is only fantasy, so in fact we had to wait until the modern evolutionary synthesis found consistency between natural selection and Mendelian genetics. Mendel's work gave natural selection its *modus operandi* and so added *gravitas*. But what if genetics could provide evidence for Lamarckian inheritance? First, it would have to show that hereditary information can also move in reverse, from body cells to genes, to enable acquisition of inheritable traits within a lifetime. Sounds controversial doesn't it? And it is; it leaps the 'Weismann barrier', the hurdle that blocks genetic information from being passed from soma to germ plasm, and so on to the next generation. This barrier was erected by the germ plasm theory of pivotal German biologist Friedrich Leopold August Weismann (1834–1914). However, somatic hypermutation, retroposons, epigenetics and horizontal gene transfer[7] are emerging as possible mechanisms to vary gene expression, move genetic material against the gene flow, and jump these very barriers set up by Darwinian evolution. However, it is worth noting that even after genes are introduced by such mechanisms, they're still susceptible to mutation; neo-Lamarckism may have been proposed as a new discipline to reunite Darwinism and a reinvigorated Lamarckism, but natural selection has always been the moderating unifier.

It is ironic that 40 years after its publication, the *Origin of Species* was considered not Darwinian enough, mainly because it adopted Lamarckian ideas; in Chapter V, 'Laws of Variation', Darwin hedges his bets with: 'The eyes of moles and of some burrowing rodents are rudimentary in size … probably due to gradual reduction from disuse, but aided perhaps by natural selection'. George Romanes (1848–1894) introduced the term 'neo-Darwinism' to classify an evolution driven solely by natural selection[8] while being restrained by the Weismann barrier and functionally comparable to DNA-based genetic heredity. It took a further 40-plus years for the modern evolutionary synthesis, coined by Julian Huxley (1887–1975) for his book *Evolution: The Modern Synthesis*, to integrate natural selection with Mendelian genetics, drawing on the new field of population genetics largely developed by Ronald A. Fisher (1890–1962), J.B.S. Haldane (1892–1964), and Sewall Wright (1889–1988). Their initial exposition had been in complex mathematical terms and was subsequently advanced and disseminated more widely, principally by Theodosius Dobzhansky (1900–1975) and Ernst W. Mayr (1904–2005). But that was over half a century ago. I asked Nick Barton about these developments, what had been happening since, and if the next revolution wasn't overdue:

> I did a degree in Natural Sciences at Cambridge and started out wanting to do physics but gradually veered towards biology mostly because the biology lectures seemed much more interesting. I think what led me into biology was the attraction of classical genetics initially and so I specialised in genetics in my final year and did a genetics degree, maybe one of the last genetics degrees; the field's changed very much. This was in 1976, which was before DNA sequencing and, therefore, when genetics had to be done in the old way, by making crossovers and by making rather elaborate inferences which gave it a certain elegance and a certain generality. It was only there really that I started to engage with evolution; I hadn't done biology in school beyond 15. So, I think it was then that I read the *Origin* [*of Species*] and *Voyage of the Beagle* and so on, and read quite a bit of the old classics as well as Darwin, Dobzhansky and so on, and was really quite struck that they encapsulated arguments which are still valid and still interesting when talking about new questions, encapsulated in a very concise way which remarkably still applies despite all the advances in genetics and in molecular biology, etc. I then did a PhD on a hybrid zone between two chromosome races of grasshopper which meant that I was doing fieldwork, and at that stage I was led into evolutionary biology. So, I was led into biology through the attraction of Mendelian genetics, classical genetics, and then, from that, into evolutionary biology by trying to solve field problems, trying to understand what was going on with these grasshoppers. And I continue to be surprised that much of evolutionary biology depends on Darwin, rather than Mendel, or rather than Watson and Crick which in some ways is surprising. It's encouraging.

Darwin obviously didn't have Mendelian technology at his fingertips and it seems a great shame because he could have done a great deal with it!

> It's actually rather intriguing that if you look at his notebooks of experimental crosses in plants, you can see Mendelian ratios there, but he never took them further.

23

It seems that Mendel's work was not something that he actually engaged with and introduced into his own writings.

Yeah, it is surprising. I think it is something that is understood by historians. I don't know the ins and outs of it, but it is interesting because he was clearly extremely interested in inheritance; he made a large effort trying to understand inheritance. Why it is that he didn't pick up Mendel's ideas, why no one picked up Mendel's ideas for so long, is really a bit of a mystery.

Well, we have a firm grasp on them now, and without Mendel it would be quite a leap from those original ideas of Darwin to say Dobzhansky or Mayr 60 years later, or so, in terms of the synthesis at the time. There's obviously the whole new format of genetic thought, but at what stage do you think that maybe everything has been taken out of Darwin's hands, and new methods are coming in which explain things that he would never have realised, because he didn't have that technology at his fingertips?

Certainly if you look at population genetics then there are processes which Darwin could not have described in any detail although I think conceptually he still understood what we've come to think of as mutation, random drift, recombination. I mean, a large part of Darwin's work was identifying sources of variation both as we would now term mutation and recombination; a large part was involved in understanding sexual reproduction, particularly a lot of his experimental work in plants: okay, we would now call that recombination, but the underlying processes were being understood in a qualitative way by Darwin. I guess that random genetic drift played less of a role [...] So yes, population genetics has a large mass of theory, of experiments and so on, which concerns processes that were only barely understood, could only barely be understood before Mendelian genetics were discovered. But having said that, if you think about the key biological issues now, a lot of those depend primarily on selection, the clear argument that selection is the key process, and that much evolutionary biology actually doesn't depend on the non-selective processes identified by population genetics. Much of, for example, certain animal behaviour studies on sexual selection, studies on speciation, and so on, these are all issues that actually go right the way back to Darwin. And those depend on these processes that we now understand in great detail, but which simply provide variation; they provide a background on which selection works.

I think probably a lot of the readership would be forgiven for not having kept up with cutting edge advances in population genetics; maybe the most recent debate to be popularised was regarding individual selection and aspects of kin selection. Are those things resolved in the modern literature?

Yes, basically. I think it's interesting that you started by talking Mayr and Dobzhansky and the modern synthesis of the 1930s and 1940s, and I think they are basically the core of the subject of evolutionary biology. One difference between our view now and our view then would be a much greater emphasis on individual selection, amongst individuals, very much in the way that Darwin emphasised, and much less emphasis on group selection, species selection and so on, which were ideas which were relatively prominent right the way through to the 1960s and 1970s, so in a way the field has come back to Darwin, come back to an understanding that it's selection among individuals that's primary.

And was it this infinitesimal variation on an individual scale that, for Darwin, made it difficult to even define species boundaries? In a sense, he didn't really recognise it and he didn't really finalise a definition of species anywhere in the book 'Origin of Species', and the whole species concept has grown pretty much independent of his ideas. This is something that has come up again with Coyne and Orr's 'Speciation' book,[9] and the debate is again on species concepts and where you can actually draw lines between groups of plants and animals. What are the latest ideas from the new synthesis, and how does individual variation fit with such categorisation?

I think, although I'm not very clear on the details, that there's a strong argument about actually what Darwin believed about the nature of the species and how that changed between his earlier notebooks in the 1840s and the *Origin of Species*, the published work. My impression is that although he emphasised the arbitrariness of species boundaries later on, that was partly to emphasise the continuity of evolution and the arguments for evolution. In fact, in his earlier writings it's very clear that reproductive isolation was a key issue and that what we now call the 'biological species concept' clearly goes back well before Mayr and Dobzhansky with whom the name is associated now, but it's really forgetting history, and those issues were already very clear way back through the 19th century. People like David Starr Jordan, and so on. That's an earlier history that Coyne and Orr do quite a good job of reviving. So, I think there are actually two aspects: the nature of individual variation and this whole issue of whether evolution involves variation to a small or large effect, which is a long-running issue, and then there's the separate issue of whether species are real discrete entities or whether they grade into each other in the extent to which they are continuous. I think they're actually separate issues, they don't depend on each other.

So, species definitions can fit okay within the Darwinian idea of extreme variation and there isn't a blurring of his ideas when it comes to species boundaries?

I think you have different views. Perhaps caricaturing it between botanists and zoologists [...] botanists have always emphasised blurring of species boundaries because hybridisation is more obvious I think in the botanical world. Certainly when you look at microbial evolution and when you go beyond what's evident, then you have to have a very different idea of species, as you cultivate them in reproductive isolation and you don't have regular sexual reproduction. So ideas of 'what a species is' depend on what group you're working on. I don't think it would be fair to emphasise extreme variation, you have to emphasise the very rapid production of variation because at the time of [Darwin's] writing he thought it all depended on gradualistic variation, and so the issue of generation of variation was much more difficult for him than it is for us now. We know about particulate inheritance or Mendelian inheritance, variation of a complete pool, say of a population. So I think it's difficult to read back to Darwin's views on variation, both because he didn't understand the nature of inheritance and also because he was trying to emphasise the continuity of evolution and therefore, if you like, emphasise the blurring of species boundaries.

And his gradualism was very much in contrast to saltation.

And also the awareness that gradualism is necessary for the evolution of complex functions and that was really his emphasis – how do complex organs arise? And much of his argument really isn't about speciation in the modern sense, it's about the evolution of adaptations and divergence.

Now, today, the debate is about speciation and it seems to have moved on from gradualism to macromutations again, and brings in punctuated equilibrium and saltation: Coyne and Orr seem to lean towards macromutation as being a mechanism. They also seem to think that even within an allopatric framework you'll get some gene flow and that doesn't negate allopatric speciation.

There are two separate issues again: one is parapatric versus allopatric speciation, in other words whether a more or less continuously distributed population can separate into distinct reproductive species, and I think that both empirically, as Coyne and Orr set out, and theoretically, it's quite clear that that can happen. Most of those mechanisms that work in allopatry, with absolute geographic isolation giving a diverging population, would also work in parapatry, in other words in a broadly continuous distribution. I think it's interesting to go back to Darwin and to see that argument still going on within his own writings and in fact in the *Voyage of the Beagle* he discusses divergence and variance across South America, and how it is that different species of *Rhea* can apparently evolve there.

The other point: it's sort of interesting you say that there's more of an interest in macromutation now. I think the definition as it is in Coyne and Orr's book, they just point out that the real evidence that we have about the genetic basis of species differences suggests that there are genes with quite a large effect, and that is what we're trying to work on identifying quantitatively using QTL [quantitative trait locus] now, and we can also see that there are some species differences which are relatively large in effect. However, there are also those with a small effect. So, their relative contributions are not clear. But there's a whole lot of different issues all tangled up; punctuated equilibrium is a question of relative rates of evolution seen in the fossil record and doesn't directly say anything about the size of genetic effect involved. The fastest change that we've seen in the fossil record is relatively slow by the standards of modern artificial selection, and so there's no particular inference that any particularly large or small effects were involved. And then there's the whole evodevo [evolutionary developmental biology] arguments where people are very interested in genes with major effects in development, and of course there are large effects that produce a fruit fly with legs and antennae, but those aren't going to go anywhere in evolutionary terms.

It does seem that those HOX genes are involved in determining variation within populations, between species, and so on, but those variations are actually alleles with a very minor effect: there's a slight difference in the binding of transcription factors which slightly tweak and tune the way those key developmental genes work. So, it's unclear what we should call it, genes with a major or minor effect. A gene can have a variety of effects; the individual alleles, the individual variance of those genes that work in evolution, may have a small effect, may have a large effect.

Creationists claim HOX genes as being a good example of a Divine hand because of their continuity across such a huge range of species.

Well, it's an extraordinary observation! But, included with the most extraordinary things to come out of molecular biology in evolutionary terms, there's been so much confirmation that living organisms are so similar. In a way it's remarkable that Darwin was confident that all living organisms had a single origin, when way back in Victorian times, their similarity was very much less clear. Now that we know that the basic molecular biology machinery, the genetic code, and all that is associated with it, are basically the same in all living organisms, that's the really striking part that comes out of molecular biology, but in itself doesn't say anything about an evolutionary mechanism. There are arguments about that the particular way in which organisms are built is particularly favourable to evolution. It's very hard to grapple with the evidence because we just have the one instance: we know that this is the way organisms are, and they evolved very successfully. Whether, if they were built in a different way, they would evolve in a different way, or adapt less well, or produce less diversity, etc., is very hard to answer.

So, you mentioned evolutionary mechanisms. We have genetic gradualism from Darwin, we have drift from Mayr and Dobzhansky. What have we discovered since then? Are we in a 'new' new synthesis at the moment, with new concepts coming in, or are we seeing a healthy, gradual progression of the field?

Actually, it's interesting, but I think there are very few new concepts and even if you look at people who, in their papers, say this is the 'new' new synthesis, you might trace those papers back to ideas that were in the 1950s, that's with [Conrad] Waddington, particularly ideas in evolutionary development, and back before that. There clearly is a vast amount of new material for evolutionary biologists to try to understand at the molecular level, and extraordinary examples of molecular adaptations and molecular conflicts of various kinds, genetic parasites for example, transposable elements, etc., etc., but those are pretty much explained, and are being explained, by all of the processes that we knew about in the 1930s and 1940s; and if we look at the understanding of DNA sequence variation, which is perhaps one of the biggest areas now of population genetics, that really does actually rest on ideas that were framed by Wright, Fisher, Haldane, developed by [Motoo] Kimura in diffusion theory, and so on, and developed more recently in terms of coalescent theory, the idea that we can understand the pattern of ancestry of a sample of genes taken from a population, and in fact those ideas of coalescence go back to ideas of identity by descent which were developed by Sewall Wright, largely.[10] So there's a great continuity in theoretical explanation even though the things we're trying to explain are much more diverse and much more remarkable than the things they had to explain 20 years ago.

So, the most important contemporary ideas are in fact not very contemporary at all! At the heart of it we have the new synthesis of Dobzhansky, Mayr and Wright, and they were all very much neo-Darwinists, in the true sense, while there now seems to be … not a doubt of Darwinism, but certainly an ongoing re-evaluation, a Lamarckian revision.

It's not an issue in evolutionary biology.

Perhaps it's a misconception also arising from the liberal application and misuse of natural selection.

You could say that's of course a problem that goes right the way back; Herbert Spencer whose Social Darwinism gave Darwinism a bad name, and greatly infuriated Darwin. You know, Darwin's ideas have always been so accessible, so fundamental, that they've been taken up and used in all sorts of very strange ways.

We would also like to find out more about significant developments, including your own work on hybrid zones.

Clearly, because species have evolved from a common ancestor, they must go through a phase in which there was partial gene flow between them, and in fact it is a remarkably common observation, that typically what we think of as a good clear-cut species actually hybridises a sister species, or something more distantly related. Sometimes that will be a low level of hybridisation across a wide area; for example, perhaps one of the best understood systems is *Drosophila pseudoobscura* and *Drosophila persimilis*, identified by Dobzhansky. He showed that there was a very, very low rate of hybridisation between those in nature, perhaps one in many thousands, at most, of matings. But remember, if we look at the ancestry of samples of genes, those show the trace of past hybridisation events: a very low level of cross-mating will allow a gene to get from one species to another and so, perhaps, most genes show a clear-cut phylogeny, which corresponds to the *pseudoobscura* and *persimilis* clear-cut species.

But some don't, and some show a much more scrambled relationship suggesting there has been some gene exchange in the past. So, that's an example where although we have a very, very distinct biologically different species, there's enough gene flow between the two to allow some genetic exchange, and perhaps to allow favourable alleles to get from one species to another so-called species.

I have tended to study a slightly different situation in which hybridisation is confined to a very narrow band, so there are many examples where the distribution of a group of species or a species is divided into a kind of mosaic separated by sharp boundaries. Perhaps the most striking example is in the *Heliconius* butterflies in South America where there are different populations which have one single warning pattern which warns predators of their distastefulness, but there are different warning colour patterns in different places, and these are separated by very narrow boundaries of the order of only a few kilometres wide, in which you find a whole mixture of varied butterflies produced by hybridisation, which are not recognised as distasteful by predators.

And there are many other examples, particularly with chromosomal differences: if you look at shrews living in Britain, they're divided into tens, perhaps hundreds, of little predator races. Very large predator races, distinguished by fusions between different chromosomes which, again, are separated by very narrow boundaries. These you can trace to only a few kilometres wide: up the Thames Valley, up near Oxford, and it goes further north, dividing the two chromosome races in Britain. And there are many examples like this of all sorts of characteristics, and they're hardly surprising; given that species have evolved from common ancestors, there must be a stage at which there is some gene flow between them. They're very good natural laboratories for understanding the genetic basis of species differences: how the genes have evolved, what effect those genes have in reducing gene exchange between the species, and whether the residual gene exchange between them in any way reduces or inhibits divergence.

And a relatively new term 'reinforcement'; could you explain a bit about that, and how it adds to the general picture?

Reinforcement actually goes back quite a long way to Alfred Russel Wallace who was in some ways more Darwinian than Darwin. He was very keen to explain absolutely every feature of the natural world in terms of selection, and in particular speciation was therefore a very large problem for Wallace. It was more of a problem than for Darwin in a way, because Wallace wanted to find some way in which natural selection could favour what he called 'hybrid sterility', and he came up with some very convoluted arguments which basically didn't work. There was a long, rather amusing correspondence between Darwin and Wallace in which Wallace tried to explain his scheme in a long series of logical points, and Darwin could not make head or tail of it. Basically, they don't work because there's no way in which selection can favour a further reduction in the fitness of hybrids. What it can do is, of course, favour a reduction in the rate of cross-mating if the two populations, or two incipient species, produce hybrids that are less fit – then there's clearly strong selection to reduce that cross-mating. And that process, of what is now called 'reinforcement', primarily, was something that was popularised by Dobzhansky who really argued very strongly for its importance as a source of reproductive isolation. And, again, I think the motivation there was to bring the role of natural selection into speciation.

In some ways it's helpful towards the origin of species; I mean, do species originate through the positive action of natural selection, or are they a side effect of other things that are going on – which would obviously be the consensus view? I think now it seems that the evidence for reinforcement is fairly strong in examples where species already overlap over a large area.

So, if like the *Drosophila pseudoobscura–persimilis* example, you have species which are overlapping over a large area, which have distinct ecological niches, fairly strong mating barriers already which allows them to coexist, then there's going to be very strong selection over a large area to perfect that isolation, both ecologically and in terms of cross-mating. So, there are some good examples where clearly assortative mating is stronger – mating barriers are stronger in areas where species overlap and reinforcement is selected.

There's really very little evidence, almost no evidence, that reinforcement is effective in narrow hybrid zones, because there selection is confined to a small area and theoretically one would expect it to have more difficulty in establishing species differences. So, it's actually unclear how important reinforcement has been in the origin of species. In a sense, it's central to the arguments about sympatric speciation; sympatric speciation models rely on reinforcement strengthening some incipient selection against intermediates. And, again, it's quite unclear if that is an important or widespread process. There are very few examples.

The problem with sympatric speciation seems to be that it is impossible to pinpoint.

Yes, it's very hard to find evidence for, so it's always possible to argue that it is actually very widespread, but we have very little positive evidence for it.

So, having established the correct historical context, where are we going now? What is the next synthesis?

I think actually at the moment, rather unfortunately ... well ... I haven't thought of what the opposite of 'synthesis' is! In the sense that there are a lot of very separate fields, and it's rather strange, from a sociological point of view, that evolutionary biology is very, very fragmented. There's a lot of talk of evolution, in lots of different areas.

Do essential concepts survive this fractionation of science?

Concepts do get across, but there's remarkably little communication between the fields. At the moment I'm interested in evolutionary computation which is apparently a very large field; a lot of people from engineering, informatics backgrounds are very interested in the idea that they can compete different algorithms or different computer programs against each other and mutate them and select them and recombine them, and so on, so as to reproduce adaptation in the way that organisms adapt. And this is an idea that goes back at least to John Holland who wrote a book in the 1970s on genetic algorithms[11] which were basically modelled on population genetic systems.

The basic idea is that you have a string of bits, let's say, which represents functions; an algorithm has to do some fitness measure, which illustrates how effective is this algorithm in its task, and you simply allow those algorithms that do better to produce more offspring, and you mutate and recombine the offspring in various ways.

Remarkable things can result, things that were quite unexpected, that do things in ingenious and original ways. There's one example which is quite interesting where the task that was given to the algorithm was to produce an effective frequency which let high frequencies pass and low frequencies were blocked, and it did this simply by combining inductors, capacitors and resistors in various ways: there was no restriction on how these could be combined. And various different ingenious circuits were evolved which had actually been patented in the past; so, these were unexpected and original, ingenious things produced by selection.

So there's lots of interest from a practical point of view. But, what's remarkable is that although there's a reasonably large literature on the theories of evolutionary computation, that's entirely separate from population genetics and there are theorems in there which can identify with population genetics, but it's just strange that these different areas can have developed over 30 years, independently.

5

SCOTTISH GEOLOGY

How slow ye move, ye heavy hours

The rate of change has always been central to evolutionary theory: slow gradualism versus abrupt saltation (from the Latin *saltus* meaning 'leap') in its several forms. Darwin's own understanding in gradualism came out of his appreciation of geological time and earth's grand old age, and again Edinburgh features large in the history of this revelation. Edinburgh-born James Hutton was fascinated by local rock formations encountered on travels throughout Scotland, to Galloway, Arran, Jedburgh, the Cairngorms, famously at Siccar Point near Berwick-upon-Tweed, and also right within the capital, in Salisbury Crags on Arthur's Seat. Hutton just couldn't connect what he was seeing with Abraham Werner's widely accepted Neptunism (Neptune, god of water and the sea in Roman mythology) which held that rocks had sedimented out of ancient flood waters, that some Neptunists, such as Robert Jameson, interpreted to be the biblical flood.

Instead, Hutton proposed a naturalistic Plutonist (Pluto, god of the underworld) argument that rock was produced from a molten earth's interior and manipulated and mixed and eroded over a great period by forces, into the formations that he had observed. Not least, he surmised that those processes must be ongoing, giving rise to an attitude of uniformitarianism, the constancy of processes. Hutton's friend John Playfair, and companion to Siccar Point, recalled that 'the mind seemed to grow giddy by looking so far into the abyss of time'. Suddenly the Young Earth based on biblical time and catastrophically sudden developments was insufficient to accommodate the evidence for these gradual processes, and the vital concept of 'deep time' was born, or at least reborn, given that the Chinese polymath Shen Kuo (1031–1095) had first recognised the gradual process of sedimentation in the 11th century.

Darwin absorbed Hutton's influence, especially regarding the continuity of nature and the true antiquity of the earth. His own formal geological training did little to instruct him in the core concepts. Attending Robert Jameson's courses during the second year in Edinburgh he should have been stimulated by lessons on stratigraphic geology, learning about the studies of Georges Cuvier (1769–1832) with mineralogist Alexandre Brongniart on the geology around Paris, which formed a direct precursor to the ageing of rocks and the dating of fossils. Instead, he found the lectures so 'dull' that '[t]he sole effect they produced on me was the determination never as long as I lived to read a book on Geology, or in any way to study the science'. In later years, he referred to his Neptunist professor as 'that old brown, dry stick Jameson', and reflected that he

> heard Professor Jameson, in a field lecture at Salisbury Craigs, discoursing on a trap-dyke, with amygdaloidal margins and the strata indurated on each side, with volcanic rocks all around us, and say that it was a fissure filled with sediment from above, adding with a sneer that there were men who maintained that it had been injected from beneath in a molten condition. When I think of this lecture, I do not wonder that I determined never to attend to Geology.

Although Jameson's course did nothing to inspire the young man, perhaps it did seed in Darwin a title for the *Origin of Species*,[1] and it also brought immediate benefits in giving him a chance to help with the collections in the University Museum, where he became an assistant and was awarded desk space. Tending the collections would have provided a flashback to his indiscriminate collecting as a childhood hobby, a status not radically different from that of the post-pubescent scientific amateurs who dominated British geology throughout this 'heroic age', until the founding of the British Geological Survey in 1835. The Survey formally made professionals of these gentlemen geologists. Jameson himself was a doctor by training who had been inspired into geology by John Walker (1730–1803), his Professor of Natural History at the University of Edinburgh from 1779 until 1803, and previously 30 years a Presbyterian parish priest – perhaps understandable, then, Jameson's biblical interpretation of Werner's flood, although, years later, he was to change his outlook in favour of Hutton.

Hutton's geological 'deep time', popularised by Lyell, and which could accommodate gradual evolution by natural selection, was hotly contested by William Thomson, later 1st Baron Kelvin (1824–1907), a mathematician and engineer at Glasgow University. Thomson had contradicted uniformitarianism by using thermodynamics to calculate the age of the sun, and the subsequent cooling time of the earth, to be insufficient for evolution to have occurred: 'I may say, strenuous on this point, that the age of the earth is definite. We do not say whether it is twenty million years or more, or less, but we say it is not indefinite. And we can say very definitely that it is not an inconceivably great number of millions of years… This earth, certainly a moderate number of millions of years ago, was a red-hot globe'. Unfortunately, Thomson had omitted nuclear fusion and radioactivity from his calculations, and so the potential for a star to burn bright for billions of years. The capacity of the earth to generate its own heat, thereby appearing younger than reality, had escaped him, but understandably so, as this was not known until the early 20th century.

Meanwhile, in the absence of this knowledge, Darwin was concerned by Thomson's calculations inasmuch as he estimated a longer period necessary for evolution before the Cambrian than afterwards. This much he confessed to James Croll (1821–1890), self-taught geologist and expert on glaciers who was appointed Resident Geologist and keeper of maps and correspondence at the British Geological Survey offices in Edinburgh from 1867[2]: 'I am greatly troubled at the short duration of the world according to Sir W. Thomson, for I require for my theoretical views a very long period before the Cambrian formation'.

Thomas Henry Huxley (1825–1895), 'Darwin's bulldog' and Acting Professor of Natural History at the University of Edinburgh from 1875 to 1877, attacked Thomson, calling to doubt his information sources: 'Mathematics may be compared to a mill of exquisite workmanship, which grinds your stuff to any degree of fineness; but, nevertheless, what you get out depends on what you put in; and as the grandest mill in the world will not extract wheat flour from peas cods, so pages of formulae will not get a definite result out of loose data'. However, Darwin had little recourse other than to maintain his convictions, foremost in his own powers of deduction: '[P]re-Silurian creatures must have lived during endless ages else my views wd be wrong, which is impossible – Q.E.D.'; and then also in Lyell's methods, about which he admitted to having insufficient understanding of the appropriate mathematics, 'which profound ignorance gives, but I cannot help observing that when applied to uncertain subjects, such as geology, it gives as uncertain results as geologists arrive at by other means; for instance, how Thomson and others differ about the thickness of the crust of the earth and the rate of cooling'. But, on this subject, Croll *was* qualified to comment:

> I think that you may quite fairly assume a very long period before the Cambrian formation, even according to Sir William Thomson's theory; for, supposing the earth to have originally been in a molten condition, a solid crust would very rapidly form, and if this crust would not break up and sink, the globe at the surface would be cool and suitable for life, although a short way down below the surface the heat was intense. This results from the slow rate at which the crust is able to conduct the heat from within.

Thanking Croll ('It is consolatory to me that you are inclined to give a little more age to the world'), Darwin reveals that Croll's contributions helped him to maintain confidence in his work, and reinforce his armour against his other critics.

6

ROCKY GROUNDS

For sair contention I maun bear;
they hate, revile, and scorn me

Fleeming Jenkin (1833–1885), Professor of Engineering at the University of Edinburgh, adopted his friend and colleague Thomson's argument that the age of the earth offered Darwinian gradualism insufficient time. But this criticism is rarely considered to have been as damaging as Jenkin's 'swamping argument'. In his anonymously penned review of the *Origin of Species*, Jenkin suggested that natural selection could not work to spread a newly introduced, inheritable trait through a population, as backcrossing would dilute its 'vigour' by half whenever mixed with the more predominant, parental type.[1] As part of Darwin's subscription to pangenesis, he considered, along with the majority, that the strength with which traits exert their effects is determined by their history of inheritance. This concept of blending inheritance is often described in terms of mixing together paints of contrasting colours, where each additional blending with another paint will further dilute a particular colour, and that once mixed, the individual chromatic components cannot be retrieved.

How troubled Darwin was by Jenkin's swamping argument is unclear. In their correspondence he congratulates Jenkin for his review of the *Origin of Species*, being 'the wittiest and in some respects the best written', while admitting to the clergyman, academic and author Charles Kingsley (1819–1875), who alerted him to it, that it 'seems to me one of the most telling Reviews of the hostile kind', and after discovering the identity of the author, to Joseph Hooker that 'Fleeming Jenkin has given me much trouble, but has been of more real use to me than any other essay or review'.

Apparently, Darwinian evolution had been successfully challenged and its central tenet of natural selection undermined. Jenkin's biography boasted about '[h]is paper on Darwin, which had the merit of convincing on one point the philosopher himself'. Indeed, Francis Darwin confirmed that '[t]he most important alterations [to the fifth edition of the *Origin of Species*] were suggested by a remarkable paper in the North British Review (June, 1867) written by the late Fleeming Jenkin', and himself commented that '[i]t is not a little remarkable that the criticisms, which my father, as I before, felt to be the most valuable ever made on his views should have come, not from a professed naturalist but from a Professor of Engineering'.[2]

However, a more considered reading of Jenkin's review and of Darwin's response reveals no such weakening of evolution by natural selection, but rather a strengthening of Darwin's position. Stephen J. Gould argued that the often misconstrued response by Darwin was to concede the error and retreat to the more established explanation of inheritance by Lamarckian mechanisms.[3] Instead, it seems that Darwin was able to use Jenkin's analysis to refine and improve his own published account of blending inheritance.

Critical to this understanding is the distinction between what were commonly termed 'individual differences' and 'single variations'. Individual differences were thought to be continuously occurring, small-scale dissimilarities: the things that make individuals individual. These differences were subsequently named 'fluctuating variation' by plant geneticist Hugo de Vries (1848–1935). In contrast, single variations are sporadically occurring, large-scale changes caused through mutation.

Jenkin had unwittingly referred to single variations alone in his critique of natural selection. Therefore, while working on the fifth edition of the *Origin of Species*, Darwin informs Wallace: 'I always thought individual differences more important than single variations, but now I have come to the conclusion that they are of paramount importance, and in this I believe I agree with you. Fleeming Jenkin's arguments have convinced me'. Perplexed, Wallace fired back: 'Dear Darwin, Will you tell me where are Fleeming Jenkin's arguments on the importance of single variations. Because I at present hold most strongly the contrary opinion, that it is the individual differences or general variability of species that enables them to become modified and adapted to new conditions'. Suitably chastised for his slovenly grammar, Darwin clarified a few days later:

> I must have expressed myself atrociously; I meant to say exactly the reverse of what you have understood. F. Jenkin argued in N[orth] Brit[ish] R[eview] against single variations ever being perpetuated and has convinced me, though not in quite so broad a manner as here put. I always thought individual differences more important, but I was blind and thought that single variations might be preserved much oftener than I now see is possible or probable. I mentioned this in my former note merely because I believed that you had come to a similar conclusion, and I like much to be in accord with

you. I believe I was mainly deceived by single variations offering such simple illustrations, as when man selects.

So, Darwin was never comfortable with the inheritance and spread of large-scale mutations that produced significant modifications, but, knowing that they both did occur, he had been obliged to consider the contribution to variation of small- and large-scale changes. Despite that, ever the gradualist, his preference was for incessant, small-scale adaptations, that could be preserved even with blending inheritance. He knew that under the correct conditions, the favourable trait would either continue to manifest itself in new individuals, or assortative mating, mating between like individuals, would conspire to minimise its dilution. Natural selection would then do the rest in establishing the trait within the population.

Having been freed of his double-edged shackle, Darwin was now able to rewrite a section in Chapter IV on 'Natural Selection' for the fifth edition of his *Origin of Species*:

[T]he preservation in a state of nature of any occasional deviation of structure, such as a monstrosity, would be a rare event; and [...] if preserved, it would generally be lost by subsequent intercrossing with ordinary individuals. Nevertheless, until reading an able and valuable article in the 'North British Review' (1867), I did not appreciate how rarely single variations, whether slight or strongly-marked, could be perpetuated.

Jenkin had actually made a minor error in his calculations, assuming half the offspring actually required to maintain a static population, which would conclude in the trait dying out through dilution; Darwin had spotted the mistake and corrected the estimates in his rewrite, some time before Arthur Sladen Davis, the assistant mathematics master at Leeds Grammar School, published a reworking in 1871. The consequence is that any introductions do not die out, but instead rapidly increase until spread throughout the population, but Darwin had not followed this through and, having corrected Jenkin's arithmetic, instead happily accepted Jenkin's original conclusion. It didn't matter. Even at the scale of the single gene, Jenkin's objections were immediately proved null and void by the rediscovery of Mendelian genetics, particulate inheritance and the modern evolutionary synthesis. It turned out that inheritable characters are passed to progeny as integral packets of information, best thought of as coloured beads. There is no dilution, and at any point it is still possible to separate out the mixture into beads of each colour.

Another anonymous review caused Darwin far fewer conceptual problems, but it ought to be mentioned, not least because it came from an even more unlikely source, a fellow naturalist and close friend. The early career paths of Richard Owen (1804–1892) and Charles Darwin were not identical, but nor were they too dissimilar. Additionally, in many other ways they could profess to having quite a lot in common. Like Darwin, Owen was born into a wealthy family, his father making a fortune as a West India merchant. Also like Darwin, Owen received his early education in England, albeit at a renowned grammar school, rather than a renowned public school. In managing to complete his schooling, he was already 20, and therefore four years Darwin's senior, when he also headed northwards to study medicine at the University of Edinburgh. Owen enrolled in 1824, just a couple of years before Darwin, and already the teaching standards were on the decline. He was so dissatisfied with the quality of teaching, especially in comparative anatomy, that, like Darwin's elder brother Erasmus after him, Owen enrolled in the Barclay School. But their paths did not cross, as he had moved to London and qualified by the time the Darwin brothers had arrived in Edinburgh.

Their first meeting was years later, at a dinner arranged by Lyell just four weeks after Darwin's return from the *Beagle* in 1836. Shortly after, Darwin began attending Owen's new series of Hunterian Lectures, along with the public and royalty. Meanwhile, Owen applied his superb expertise in comparative anatomy to Darwin's South American fossils, showing the link between extinct rodents and sloths to living South American species. Grant also offered to help examine some of the *Beagle* specimens. Coincidental with Darwin's departure from Edinburgh, and Knox's implication in Burke and Hare's murders, Grant had accepted, in 1827, a position as Professor of Comparative Anatomy at University College London. Now in London, he was in a position to provide much-needed assistance. However, about this time Owen launched a long and bitter campaign against Grant, mostly for his transmutationist views, but also ostensibly motivated through jealousy and ambition, character traits that would also turn him against Darwin in response to the *Origin of Species*.

Owen wanted to replace Grant as the foremost comparative anatomist, and he opened by getting him voted out of his place on the Zoological Society council which Owen then swiftly occupied. Owen's protracted deviousness isolated Grant from professional circles and this ostracism likely influenced Darwin's rejection of assistance that he would have welcomed had it not been for the controversy surrounding Grant. Further, unfortunate circumstances drove Grant into financial ruin; these included a change in the funding of university teachers and a requirement to be examined by the College of Physicians in order to practise, something to which his pride could not allow him to agree, his being already a Fellow of the Royal College of Physicians of Edinburgh. Being a radical, he was ultimately holding out for a reformation of medical licensing laws, but by 1849, 'great Grant' was discovered living in a London slum, about which he commented: 'I have found the world to be chiefly composed of knaves and harlots, and I would as lief live among the one as the other'. Remarkably, throughout his misfortune he stayed in the same, poorly paid job, for 47 years, until his death.

When Owen attacked transmutation as Lamarckian heresy, Grant lacked the volume of support he needed to defend his position. By leaving Scotland, he had exposed himself to a professional community of comparatively close-minded individuals. It was also mooted that he was homosexual, considered to be a heinous crime in Victorian Britain, capable of causing mass hysteria when brought into the open, such as during the trial of Oscar Wilde. However, there is no proof of Grant's sexual orientation, and so we cannot interpret his motivations in life in that light, nor cast his friendship with the younger Darwin in terms of 'the love that dare not speak its name'.[4]

Although Owen would often involve himself in feuds, he also managed a succession of exemplary investigations, published over 600 papers, and made major contributions towards the study of the animal kingdom – for example, classification and naming of the *Dinosauria*, descriptions of the Pearly Nautilus and *Archaeopteryx*, postulation of the Giant Moa, and introduction of homology. In 1856, in recognition of his anatomical and curating skills, he was made first Superintendent of the Natural History Collections at the British Museum, but his long-term goal was for a 'cathedral to nature', that could comfortably house the natural history specimens, distinct from the rest of the British Museum. He realised that dream in 1881 when the Natural History Museum opened in London, for which he was knighted, and he then retired, to lead a peaceful end to his life.

Throughout his early career, Owen's more honourable achievements cultivated a superb reputation for comparative anatomy which led to him becoming the obvious destination for London Zoo's deceased members of their menagerie, and for a large number of newly discovered species sent home by Victoria's intrepid explorers (often to Owen's own home, to the exasperation of his wife!). Consequentially, Owen was first to describe the anatomy of the gorilla, but on this occasion his personal beliefs biased his science (though it is sometimes difficult to know what those beliefs were in actuality). Owen was clearly a complex character. An exceptionally gifted scientist, he allowed ambition to cloud his clarity of mind, such that on occasions he was forced to manipulate his findings to suit.

In the case of the gorilla, Owen would have known that if any organs were found homologous with humans, that would be evidence for descent – a concept already being debated fiercely, prior to Darwin's contribution proper. A major reason for this was the publication of *Vestiges of the Natural History of Creation* in 1844. In the wake of Lamarck, it proposed a similar transmutational path of increasing complexity, from apes to humans, and so, in fear of the blasphemy laws, was anonymously published by the Scottish publisher and gentleman geologist Robert Chambers (1802–1871). The inevitable controversy caused a scandal, and the book was an international bestseller for the next decade. Some of that controversy came from the lambasting Chambers got, mainly for his amateur geology. Darwin read the book and learnt from the criticisms, knowing full well that his own work would attract similar attention. As part of his response to Chambers, Owen produced a couple of papers between 1853 and 1855 emphasising the differences between humans and apes and the impossibility of shared origins. Another paper followed in 1858 which formalised these differences via a new classification system for mammals.

Owen was arguing from a bias in favour of anti-materialism.[5] When it came to the gorilla, Owen's prejudices led to a line of reasoning around an area of the brain called the *Hippocampus minor*. He claimed gorillas, and other apes, lacked this and two other structures that are present in human brains, showing that humans could not have evolved from apes, concluding: 'Thus [man] fulfills his destiny as the supreme master of this earth and of the lower creation'. Owen resurrected this argument that apes lacked features of the human brain in a debate with Huxley in Oxford in 1860. Huxley simply denied that it was so. In a series of lectures between 1860 and 1863, primarily aimed at the general public, and culminating in *Evidence as to Man's Place in Nature*, Huxley repeatedly disproved these claims, irreparably damaging Owen's scientific credibility as a result. The debate was prominent enough to be immortalised in fiction. Kingsley satirised the dispute in *The Water-Babies*:

> You may think that there are other more important differences between you and an ape, such as being able to speak, and make machines, and know right from wrong, and say your prayers, and other little matters of that kind; but that is a child's fancy, my dear. Nothing is to be depended on but the great hippopotamus test. If you have a hippopotamus major in your brain, you are no ape, though you had four hands, no feet, and were more apish than the apes of all aperies. But if a hippopotamus major is ever discovered in one single ape's brain, nothing will save your great- great- great- great- great- great- great- great- great- great- greater- greatest- grandmother from having been an ape too.

And in May 1861 *Punch* published the following verse, along with the accompanying caricature:

Monkeyana

by *GORILLA*

Zoological Gardens, May, 1861

Am I satyr or man?
Pray tell me who can,
And settle my place in
the scale.
A man in ape's shape,
An anthropoid ape,
Or monkey deprived of
his tail?

The *Vestiges* taught,
That all came from
naught
By "development," so
called, "progressive;"
That insects and worms
Assume higher forms
By modification
excessive.

Then Darwin set forth
In a book of much
worth,
The importance of
"nature's selection;"
How the struggle for life
Is a laudable strife,
And results in "specific
distinction."

Let pigeons and doves
Select their own loves,
And grant them a
million of ages,
Then doubtless you'll
find
They've altered their
kind,
And changed into
prophets and sages.

Leonard Horner relates,
That Biblical dates
The age of the world
cannot trace;
That Bible tradition,
By Nile's deposition,
Is put to the right about
face.

Then there's Pengelly
Who next will tell ye
That he and his
colleagues of late
Find celts and shaped
stones
Mixed up with cave
bones
Of contemporaneous
date.

Then Prestwich, he pelts
With hammers and celts
All who do not believe
his relation,
That the tools he
exhumes
From gravelly tombs
Date before the Mosaic
creation.

Then Huxley and Owen,
With rivalry glowing,
With pen and ink rush to
the scratch;
'Tis Brain *versus* Brain,
Till one of them's slain,
By JOVE! it will be a
good match!

Says Owen, you can see
The brain of Chimpanzee
Is always exceedingly small,
With the hindermost "horn"
Of extremity shorn,
And no "Hippocampus" at all.

The Professor then tells 'em,
That man's "cerebellum,"
From a vertical point you can't see;
That each "convolution"
Contains a solution
Of "Archencephalic" degree.

That apes have no nose,
And thumbs for great toes,
And a pelvis both narrow and slight;
They can't stand upright,
Unless to show fight,
With 'Du Chaillu,' that chivalrous knight!

Next Huxley replies,
That Owen he lies,
And garbles his Latin quotation;
That his facts are not new,
His mistakes not a few,
Detrimental to his reputation.

"To twice slay the slain,
By dint of the Brain,
(Thus Huxley concludes his review)
Is but labour in vain,
Unproductive of gain,
And so I shall bid you 'Adieu'!"

One of the most damning accusations that can be made of a scientist is plagiarism. By the time the *Origin of Species* was published, Owen already had been caught falsely claiming the discovery of a type of 'belemnite', an extinct group of molluscs closely related to modern squids. This serious offence cost Owen his places on the councils of both the Zoological Society and the Royal Society. Now, Owen was well known to be anti-evolution. So, when it came to his anonymous review in the *Edinburgh Review* of the *Origin of Species* in 1860, in which he suggested, in the third person, to have usurped Darwinian evolution by a full 10 years, the Darwinians were understandably outraged, particularly Huxley, Owen's despised adversary over human origins. In contrast to his dealings with the abandoned Grant, when Owen attacked Darwinian evolution as presented in the *Origin of Species*, he found himself confronted by what essentially constituted a team of Darwin's supporters. His ruse hadn't worked, even though, for the purposes of subterfuge, Owen had fiendishly incorporated three of his own recent works, which he then proceeded to review, but was actually attempting to present evidence of an ongoing development in his evolutionary ideas:

> In his last published work Professor Owen does not hesitate to state 'that perhaps the most important and significant result of palæontological research has been the establishment of the axiom of the continuous operation of the ordained becoming of living things' [...] As to his own opinions regarding the nature or mode of that 'continuous creative operation', the Professor is silent. He gives a brief summary of the hypotheses of others, and as briefly touches upon the defects in their inductive bases. Elsewhere he has restricted himself to testing the idea of progressive transmutation by such subjects of Natural History as he might have specially in hand: as, e.g. the characters of the chimpanzee, gorilla, and some other animals.

Darwin, confused by the actions of his old friend, expresses deep hurt and humility when describing events to Asa Gray (1810–1888), his most ardent supporter in America:

> Have you seen how I have been thrashed by Owen in last Edinburgh: he misquotes & misrepresents me badly, & how he lauds himself. But the manner in which he sneers at Hooker is scandalous, to speak of his Essay & never allude to his work on Geograph. Distribution is scandalous. When Hooker's Essay appeared Owen wrote a note, which I have seen, full of strongest praise! What a strange man he is. All say his malignity is merely envy because my Book has made a little noise. How strange it is that he can be envious about a naturalist, like myself, immeasurably his inferior! But it has annoyed me a good deal to be treated thus by a friend of 25 years duration. He might have been just as severe without being so spiteful. Owen consoles himself by saying that the whole subject will be forgotten in ten years.

In confidence to Lyell, Darwin was more matter-of-fact, but still quite incredulous of Owen's departures from reality:

> I have very long interview with Owen, which perhaps you would like to hear about, but please repeat nothing. Under garb of great civility, he was inclined to be most bitter & sneering against me. Yet I infer from several expressions, that at bottom he goes immense way with us. He was quite savage & crimson at my having put his name with defenders of immutability. When I said that was my impression & that of others, for several had remarked to me, that he would be dead against me: he then spoke of his own position in science & that of all the naturalists in London, 'with your Huxleys', with a degree of arrogance I never saw approached. He said to effect that my explanation was best ever published of manner of formation of species. I said I was very glad to hear it. He took me up short, 'you must not at all suppose that I agree with it in all respects'. I said I thought it no more likely that I shd be right on nearly all points, than that I shd toss up a penny & get heads twenty times running.

I asked him which he thought the weakest parts, he said he had no particular objection to any part. He added in most sneering tone if I must criticise I shd say 'we do not want to know what Darwin believes & is convinced of, but what he can prove'. I agreed most fully & truly that I have probably greatly sinned in this line, & defended my general line of argument of inventing a theory, & seeing how many classes of facts the theory would explain. I added that I would endeavour to modify the 'believes' & 'convinceds'. He took me up short, 'You will then spoil your book, the charm of(!) it is that it is Darwin himself'. He added another objection that the book was too 'teres atque rotundus', that it explained everything & that it was improbable in highest degree that I shd succeed in this. I quite agree with this rather queer objection, & it comes to this that my book must be very bad or very good. Lastly I thanked him for Bear & Whale criticism, & said I had struck it out. 'Oh have you, well I was more struck with this than any other passage; you little know of the remarkable & essential relationship between bears & whales'.

I am to send him the reference, & by Jove I believe he thinks a sort of Bear was the grandpapa of Whales! I do not know whether I have wearied you with these details which do not repeat to any one. We parted with high terms of consideration; which on reflexion I am almost sorry for. He is the most astounding creature I ever encountered.

All Darwin could do was declare in resignation and with regret: 'The Londoners say he is mad with envy because my book is so talked about. It is painful to be hated in the intense degree with which Owen hates me'. Even in death he would continue to rile Owen on a daily basis: a marble statue to commemorate Darwin was sited on the landing in Owen's Natural History Museum. In 1927 the tables were turned and Owen's bronze was put in pride of place, while Darwin's statue was relegated to a rear annex, alongside a statue of Huxley. However, in this anniversary year, Darwin has been reinstated and Owen placed at the foot of the stairs, mere metres away and a lot closer than one could have brought them in reality after 1859. One of the last straws for Darwin, of Owen's deviousness, occurred in 1871, when Owen attempted to sabotage funding of Hooker's botanical collection at Kew Gardens. This was possibly motivated by a selfish wish to bring it under Owen's jurisdiction within the British Museum. Darwin commented: 'I used to be ashamed of hating him so much, but now I will carefully cherish my hatred & contempt to the last days of my life'.

Despite such hatred, perhaps the most notorious enemy of the Darwinists was Louis Agassiz (1807–1873), although he was quite admiring of Darwin himself, of whom he wrote in the 1869 edition of his 1857 *Essay on Classification*:

I have for Darwin all the esteem which one has to have; I know the remarkable work that he has accomplished, as much in Paleontology as in Geology, and the earnest investigations for which our science is indebted. But I consider it a duty to persist in opposition to the doctrine that today carries his name. I indeed regard this doctrine as contrary to the true methods that Natural History must inspire, as pernicious, and as fatal to progress in this science. It is not that I hold Darwin himself responsible for these troublesome consequences. In the different works of his pen, he never made allusion to the importance that his ideas could have for the point of view of classification. It is his henchmen who took hold of his theories in order to transform zoological taxonomy.

However, this was prior to setting out his religion-based disagreements for Darwinism, especially on speciation, which are still core to religion-based disagreements today: namely, Darwinism is hypothetical and fits evidence to suit, variation is limited within kind, and the fossil record is inconsistent. But, nearly 20 years before finding his *bête noire*, and even less time since Darwin had left, Agassiz visited Edinburgh, in 1840.

Like Owen, Agassiz was a ruthlessly ambitious individual, but his lack of scruples merely incited denunciation from the very ranks of great men with which he wished to be associated, sadly overshadowing his more honest achievements. On this tour he was energetically publicising an idea, from a few years before, that there had been an ice age in the northern hemisphere, and so was no doubt relieved to discover vindicatory striations in the rock and a hermetically isolated large boulder in the Hermitage of Braid on the south side of the city, the first evidence of glacial erosion, transportation and deposition in Scotland. On emigrating to America, he repeated this conclusion for the boulders strewn across New England, refuting the belief that they had been distributed by Noah's flood, where in his honour there are also the Big and Little Agassiz rocks.

Ironically, even if Darwin had first encountered glacial theory elsewhere, Agassiz was a major proponent and likely contributed to Darwin's geological convictions, if only indirectly. In return, Agassiz watched and fumed as all around him turned to Darwinism, while others, including Darwin, passed him by and yet revelled in the concepts with which he had fought so hard to be identified. Like on one occasion, recalled in Darwin's 'autobiography'[6] when he was still a 13- or 14-year-old schoolboy in Shrewsbury:

[A]n old Mr [Richard] Cotton in Shropshire, who knew a good deal about rocks, had pointed out to me two or three years previously a well-known large erratic boulder in the town of Shrewsbury, called the bell-stone; he told me that there was no rock of the same kind nearer than Cumberland or Scotland, and he solemnly assured me that the world would come to an end before anyone would be able to explain how this stone came where it now lay. This produced a deep impression on me and I meditated over this wonderful stone. So that I felt the keenest delight when I first read of the action of icebergs in transporting boulders, and I gloried in the progress of Geology.

Just before Agassiz's tour, Darwin returned to Scotland, taking in Edinburgh and going as far as the Scottish Highlands, in June 1838. His notebook from that period records:

[1838]

June — (beginning). Preparing 1st Part of Birds — St Jago geology — some little Species theory, & lost very much time by being unwell.—

— 23d Started in Steam boat to Edinburgh (one day Salisbury Craigs). Spent eight good days in Glen Roy. returned by sea through Greenock & Liverpool, slept at Overton

& reached Shrewsbury July 13th.

July 29th. Set out for Maer

August 1st. London. Began paper on Glen Roy & finished it

September 6th. Finished paper on Glen Roy — one of the most difficult & instructive tasks I was ever employed on

Sept. 14th. Frittered these foregoing days away in working on Transmutation theories & correcting Glen Roy. Began craters of Elevation Theory

This whole period, the expedition, its stopovers and the subsequent roundabout return, he had plotted in great detail from his bachelor pad at 36 Great Marlborough Street in London, revealed in a letter to his second cousin William Darwin Fox (1805–1880):

I have not been very well of late, which has suddenly determined me to leave London earlier than I had anticipated. I go by the steam-packet to Edinburgh. take a solitary walk on Salisbury crags & call up old thoughts of former times then go on to Glasgow & the great valley of Inverness, near which I intend stopping a week to geologise the parallel roads of Glen Roy, thence to Shrewsbury, Maer for one day, & London for smoke, ill health & hard work.

Some two months after returning, towards the end of this period, he was in a position to recount his adventure to Charles Lyell, greatly praising Scotland, but also with poetic wickedness and relief:

[M]y trip in the steam packet was absolutely pleasant, & I enjoyed the spectacle, wretch that I am, of two ladies & some small children quite sea sick, I being well. Moreover on my return from Glasgow to Liverpool, I triumphed in a similar manner over some full grown men. I staid one whole day in Edinburgh, or more truly on Salisbury Craigs. I want to hear, some day, what you think about that classical ground: the structure was to me new & rather curious, that is if I understand it right. I crossed from Edinburgh in gigs & carts, (& carts without springs as I never shall forget) to Loch Leven, was disappointed in the scenery & reached Glen Roy on Saturday evening, one week after leaving Marlborough St. Here I enjoyed five days of the most beautiful weather, with gorgeous sunsets, & all nature looking as happy as I felt. I wandered over the mountains in all directions & examined that most extraordinary district. I think without any exception, not even the first volcanic island, the first elevated beach, or the passage of the Cordillera, was so interesting to me, as this week. It is far the most remarkable area I ever examined. I have fully convinced myself, (after some doubting at first) that the shelves are sea-beaches, although I could not find a trace of a shell, & I think I can explain away most, if not all, the difficulties. I found a piece of a road in another valley, not hitherto observed, which is important; & I have some curious facts about erratic blocks, one of which was perched up on a peak 2200 ft above the sea. I am now employed in writing a paper on the subject, which I find very amusing work, excepting that I cannot anyhow condense it into reasonable limits. At some future day I hope to talk over some of the conclusions with you which the examination of Glen Roy has led me to. Now I have had my talk out, I am much easier, for I can assure you Glen Roy has astonished me.

However, while Darwin may have enjoyed the stimulation of the Glen Roy paper, he had also made a mistake in his analysis. Darwin misinterpreted the famous 'parallel roads' running laterally along the side of the valley as ancient shorelines. He had interpreted sedimentation and had missed the telltale signs of glaciation, and so the road to Agassiz, who no doubt would have welcomed the recognition of having been cited by Darwin. Rough justice maybe, yet most of the time Agassiz was the main reason behind his own downfall.[7] Perhaps it was therefore more than purely seismic, and a little karmic, that a statue of him was knocked from the second storey at Stanford University by the 1906 San Francisco earthquake? 'Agassiz in the concrete' was described thus: 'A big marble statue of Agassiz was toppled off his perch on the outside of the quad and fell foremost into the ground (right through a cement walk) up to his shoulders, and still sticks there, legs in the air and his hand held out gracefully. People came running from the quad with such sober faces, but when they saw him they couldn't help laughing, and one fellow went up and shook hands with him'. And where did it land? Head buried, plumb outside the Zoology Department!

Darwin knew that he had made a mistake on Glen Roy, coming to call his paper his 'one long gigantic blunder'. Even though he was embarrassed, he maintained interest so that when Edinburgh geologist David Milne published his 1847 paper 'On the parallel roads of Lochaber' which proposed that freshwater lakes had formed the parallel roads, Darwin contacted Chambers, who had also become interested in the terraces, to request a copy of the paper. A copy from Lyell reached him first, but he then asked Chambers for a meeting in Edinburgh which he never made, supposedly through bad health, but continued his discussion by correspondence with Chambers, Lyell and Milne. Darwin was eventually persuaded of Agassiz's glacial lakes, while

Chambers converted to Milne's version. Darwin then submitted a letter of reply with suggestions for further work to *The Scotsman*, but it was rejected by the newspaper's editor Charles Maclaren for it being too technical for his readers. Maclaren forwarded the letter to Jameson's *Edinburgh New Philosophical Journal* which Darwin must have objected to as he asked Jameson to destroy it. It was likely also unsuitable for a technical audience, but Darwin had probably had enough of Glen Roy by now, complaining to Hooker that '[t]he confounded subject has made me sick twice'.

Darwin would return to Scotland just once more, this time accompanied by his wife, in 1855 for a British Association meeting in Glasgow: 'I really have no news: the only thing we have done for a long time, was to go to Glasgow; but the fatigue was to me more than it was worth & Emma caught a bad cold [...] I saw a little of Sir Philip [de Malpas Grey-Egerton] (whom I liked much) [...] The meeting was a good one & [George Douglas Campbell] the Duke of Argyll spoke excellently'. Although that was his last time north of the border, Darwin would be presented with other opportunities. A medical student at the University of Aberdeen wrote to him in 1872, saying that the students wished to nominate him as Lord Rector of their university, and enquired whether his health would be an obstacle. Darwin replied that he was honoured to have been asked, but had to decline due to the 'status of my health'. Huxley took the post instead.

Part of Hutton's evidence towards a disproof of Neptunism included what is now known as 'Hutton's Section' at the base of Salisbury Crags. The rocks in that area clearly showed that molten (igneous) rock had been pushed in between existing sedimentary rocks, breaking them apart.

Photographs © Melinda T. Hough 2010

Agassiz's Rock is the large tapered boulder at the foot of Blackford Hill (background) in the Hermitage of Braid, to the south of Edinburgh. Once more in the ice and snow, it was originally identified by Louis Agassiz as an early example of ancient glacial action that had smoothed and polished its surfaces.

Photographs from author's collection

7

BEAGLE VOYAGE

Behold the hour, the boat arrive

Uninspired about medicine, Darwin was removed from Edinburgh in April 1827 and enrolled by his father to train for the clergy at Christ's College, Cambridge, where he confesses: 'During the three years which I spent at Cambridge my time was wasted, as far as the academical studies were concerned, as completely as at Edinburgh and at school'. The major difference, of course, was that Cambridge provided neither the environment nor individuals that could stimulate the naturalist in Darwin. Instead, he reverted to his sporting interests:

> Although, as we shall presently see, there were some redeeming features in my life at Cambridge, my time was sadly wasted there, and worse than wasted. From my passion for shooting and for hunting, and, when this failed, for riding across country, I got into a sporting set, including some dissipated low-minded young men. We used often to dine together in the evening, though these dinners often included men of a higher stamp, and we sometimes drank too much, with jolly singing and playing at cards afterwards. I know that I ought to feel ashamed of days and evenings thus spent, but as some of my friends were very pleasant, and we were all in the highest spirits, I cannot help looking back to these times with much pleasure.

However, Darwin did register on a natural history course taught by another gentleman geologist and professor of botany, the Reverend John Stevens Henslow (1796–1861). Unlike the University of Edinburgh which discarded its early chapel and chaplain, Cambridge University was built on Anglicanism, so it was inevitable that natural history was going to be taught as a 'natural theology', a crusade for Divine design in nature, in the vein of William Paley's dogmatic tome of the same name. Darwin didn't get on very well with ideas like those of Paley; nonetheless, the timely influence of Henslow on Darwin's future cannot be overstated. Without Henslow's intervention and interest in Darwin's career,

50

all the foundations of rationalism seeded in the young man's mind while at Edinburgh may have been lost. Having mentored him and tutored him through his finals in January 1831, Henslow guided Darwin away from immediate priesthood, refocusing his attention on the geology that had excited him whilst at Edinburgh, and introduced him to the Woodwardian Professor of Geology Adam Sedgwick (1785–1873). He also encouraged Darwin's dreams of travel, ultimately setting him up as unpaid gentleman's companion and naturalist for an intended two years aboard the second voyage of Captain Robert FitzRoy's HMS *Beagle*, a Cherokee-class Royal Navy sloop.

Darwin's voracious appetite for reading clearly stayed with him from his Edinburgh days. The wanderlust and scientific zeal were further fired for Darwin by his reading in that year, in particular, Alexander von Humboldt's *Personal Narrative* and Sir John Herschel's *Introduction to the Study of Natural Philosophy*. The first opportunity for satiation was a postgraduate trip to Tenerife for which he needed to hone his field skills. So, in August 1831, Darwin found himself back in North Wales, this time on a geological tour with Sedgwick. They worked northwards and eastwards from Llangollen, travelling on to Llanfairfechan and Penmaenmawr, Bethesda and Llanberis. Darwin may not have been there for very long, but in just three weeks he made some invaluable observations that led to useful publications and revision of existing ones; helped unearth a rhinoceros fossil; and, much to my shame, covered as much ground as I managed in six years of living there!

In his absence, events at home were also moving fast. When he returned to the family home in Shrewsbury, a life-changing letter from Henslow, addressed to 'C. Darwin Esq' Shrewsbury To be forwarded or opened, if absent', awaited:

Cambridge

24 Aug 1831

My dear Darwin,

Before I enter upon the immediate business of this letter, let us condole together upon the loss of our inestimable friend poor Ramsay of whose death you have undoubtedly heard long before this. I will not now dwell upon this painful subject as I shall hope to see you shortly fully expecting that you will eagerly catch at the offer which is likely to be made you of a trip to Terra del Fuego & home by the East Indies. I have been asked by Peacock who will read & forward this to you from

London to recommend him a naturalist as companion to Capt Fitzroy employed by Government to survey the S. extremity of America. I have stated that I consider you to be the best qualified person I know of who is likely to undertake such a situation. I state this not on the supposition of yr. being a finished Naturalist, but as amply qualified for collecting, observing, & noting any thing worthy to be noted in Natural History. Peacock has the appointment at his disposal & if he can not find a man willing to take the office, the opportunity will probably be lost. Capt. F. wants a man (I understand) more as a companion than a mere collector & would not take any one however good a Naturalist who was not recommended to him likewise as a gentleman. Particulars of salary &c I know nothing. The Voyage is to last 2 yrs. & if you take plenty of Books with you, any thing you please may be done— You will have ample opportunities at command. In short I suppose there never was a finer chance for a man of zeal & spirit. Capt F. is a young man. What I wish you to do is instantly to come to Town & consult with Peacock (at Nº. 7 Suffolk Street Pall Mall East or else at the University Club) & learn further particulars. Don't put on any modest doubts or fears about your disqualifications for I assure you I think you are the very man they are in search of— so conceive yourself to be tapped on the Shoulder by your Bum-Bailiff & affecte friend | J. S. Henslow

(Turn over)

The expedn. is to sail on 25 Sept: (at earliest) so there is no time to be lost

Darwin's Tenerife expedition was cancelled after the very recent death of his travelling companion Marmaduke Ramsay, so Henslow's letter must have provided a welcome and exciting distraction.

Another of Sedgwick's contributions to the scientific success of the *Beagle* expedition was in reply to a request for recommendations for books to take on the voyage:

[...] I cannot but be glad at your appointment & I truly hope it will be a source of happiness & honor to you. I really dont know what to say about books. Nº. 1 Daubeny. Nº. 2. a book on Geology. D'aubuissions work is one of the best tho' full of Wernerian nonsense.

I dont think Bakewell a bad book for a beginner. For fossil shells what is to be done? Go to the Geological Society and introduce yourself to Mr Lonsdale as my friend & fellow traveller & he will counsel you. Humboldts personal narrative you will of course get. He will at least show the right spirit with w^h. a man should set to work. There is a small paper printed by the Geol. Soc^y containing directions for travellers &c– Lonsdale will give you a copy: but it is a mere horn book hardly worth your looking at. Study the Geological Soc^{ys}. collection as well as you can & pay them back in specimens. I am to propose you when the meetings begin.

He would probably have suggested differently if he had known where the *Beagle* voyage was to lead; Sedgwick could never accept Darwinian evolution, as with Agassiz, for an argument still made today: it omits the metaphysics of humanity. Nonetheless, Darwin most certainly paid his dues in specimens, but while there is a well-worn copy of Daubeny on the bookshelves at Down House, which may be a well-travelled copy, the other recommendations were not necessarily taken on board, metaphorically and literally. Sedgwick must have honoured his pledge too because Darwin was made a Fellow of the Geological Society in November 1836 and elected its secretary in February 1838 – the same year of his visit to Glen Roy that produced the erroneous paper. Evidently the Geological Society didn't consider Glen Roy to be too bad a mistake as it didn't stop them awarding Darwin the Wollaston Medal, their highest accolade, in 1859 for the *Origin of Species*. But it must have stuck in the craws of two of their previous laureates, Louis Agassiz (in 1836) and Adam Sedgwick (in 1851).

Another associate of Henslow's proved to be far greater a supporter of Darwin than Sedgwick, and would become 'the one living soul from whom I have constantly received sympathy'. Joseph Dalton Hooker (1817–1911) was born in Suffolk, but raised in Glasgow. He attended lectures in botany given by his father, William Jackson Hooker, from the age of seven. Educated at the Glasgow High School, he successfully studied medicine at Glasgow University, and like Darwin, yearning for travel, upon graduating in 1839 joined a maritime expedition. It is possible that he met the already famous Darwin, briefly, in Trafalgar Square, just prior to embarkation. What is certain is that Hooker was impressed by Darwin's *Voyage of the Beagle*; Charles Lyell had given him the proofs to read during his own voyage, due to call in at some of the same places visited by Darwin.

Coincidentally, Lyell had already influenced Darwin's reading for his voyage (a decade before Agassiz had even set foot in Scotland). It was Hutton's work that fellow Scot, Charles Lyell (1797–1875), had expanded on in his *Principles of Geology*, possibly on glacial theory but definitely on uniformitarianism, that captivated Darwin's attention and informed him about his observations:

> The investigation of the geology of all the places visited was far more important, as reasoning here comes into play. On first examining a new district nothing can appear more hopeless than the chaos of rocks; but by recording the stratification and nature of the rocks and fossils at many points, always reasoning and predicting what will be found elsewhere, light soon begins to dawn on the district, and the structure of the whole becomes more or less intelligible. I had brought with me the first volume of Lyell's *Principles of Geology*, which I studied attentively; and this book was of the highest service to me in many ways. The very first place which I examined, namely St. Jago in the Cape Verde islands, showed me clearly the wonderful superiority of Lyell's manner of treating geology, compared with that of any other author, whose works I had with me or ever afterwards read.

Henslow sent Darwin a copy of Lyell's second volume of his *Principles of Geology* direct to the *Beagle* upon its publication in 1832. If Volume 1 had influenced him on geology, then Volume 2 had an altogether different and less well understood influence on Darwin's comprehension of natural processes, and specifically his conclusion that 'all living things are bound together by a web of complex relations'. This connectivity, bonded through the shared struggle for survival, generates relationships between conspecifics (members of the same species), between predators and their prey, and between all life and its physical habitat, or environment. This synthesis of biotic and abiotic is nowadays called 'ecology', first coined from the Greek (*oikos*: house and *-logia*: study) by the German polymath Ernst Haeckel (1834–1919). In fact, the first ecologists were the Ancient Greeks, notably Theophrastus (371–*c.*287 BC) and Aristotle, who recorded their observations on the interaction between an environment and its plant and animal inhabitants. Darwin's contemporaries were also making important contributions to this concept of biological communities coexisting within an ecosystem: Alexander Humboldt (1769–1859), Alfred Russel Wallace and Karl Möbius (1825–1908). But Darwin, above all, revealed the underlying complexity by showing how well plants and animals are adapted to their environments. Crucially, this has also exposed the underlying fragility of nature and its sensitivity to human activity. Without such insight we may never have seen James Lovelock's Gaia hypothesis, the Green movement nor our capacity to fully understand our impacts on earth, including anthropogenic climate change. However, while Darwin was still aboard the *Beagle*, Lyell was already delivering a distinctly environmentally friendly message in his second volume, and this undoubtedly informed Darwin's own understanding:

> [...] the various species of contemporary plants and animals have obviously their relative forces nicely balanced, and their respective tastes, passions,

and instincts, so contrived, that they are all in perfect harmony with each other. In other manner could it happen, that each species surrounded as it is by countless dangers should be enabled to maintain its ground for periods of considerable duration [...] Yet, if we wield the sword of extermination as we advance, we have no reason to repine at the havoc committed, nor to fancy, with the Scotch poet, that 'we violate the social union of nature;' or complain, with the melancholy Jaques, that we

> Are mere usurpers, tyrants, and, what's worse,
> To fright the animals, and to kill them up
> In their assign'd and native dwelling-place.

We have only to reflect, that in thus obtaining possession of the earth by conquest, and defending our acquisitions by force, we exercise no exclusive prerogative.

Thus, within the pages of his select library is one of the places where Darwin expanded his mind during his extended voyage aboard the HMS *Beagle* (from 27 December 1831 until 2 October 1836), but, like Hooker, Darwin's fascination for the natural world was also driven by his wanderlust, something that had peaked within him while still at Edinburgh. The *Beagle* voyage fulfilled those fantasies, bringing to life the other worlds he had only been able to visit through the likes of Milton's *Paradise Lost*, a copy of which he carried on board and referred to frequently. Remarkably, Darwin actually ended up spending just 39 months on land and only 18 months at sea. Even so, the voyage faced him with the reality of a world he had only imagined up until then, and when it did, it was Edinburgh to which he turned for a non-fictional comparison, whilst hardly containing his excitement:

1832

Jan. 17th

Immediately after breakfast I went with the Captain to Quail Island. This is a miserable desolate spot, less than a mile in circumference. It is intended to fix here the observatory & tents; & will of course be a sort of head quarters to us. Uninviting as its first appearance was, I do not think the impression this day has made will ever leave me. The first examining of Volcanic rocks must to a Geologist be a memorable epoch, & little less so to the naturalist is the first burst of admiration at seeing Corals growing on their native rock. Often whilst at Edinburgh, have I gazed at the little pools of water left by the tide: & from the minute corals of our own shore pictured to myself those of larger growth: little did I think how exquisite their beauty is & still less did I expect my hopes of seeing them would ever be realized.

Darwin's time with Grant was now paying dividends, although their beachcombing had ended in a rift over Darwin's discovery of voluntary movement in zoophyte eggs. Grant considered this an impingement too far on his subject area. Even though Grant made some concession by acknowledging his 'zealous young friend Mr. Charles Darwin' in the resultant paper, their friendship was never repaired. Darwin had been shocked by Grant's apparent selfishness, but only aggrieved to the extent that he felt he could still turn to Grant for advice on specimen storage. Darwin visited him in London while preparing for the *Beagle* voyage. Once aboard HMS *Beagle*, Darwin methodically applied the practical skills that he had learnt, mainly from his time in Scotland: collecting specimens, tracing larval development, dissecting marine invertebrates and investigating their anatomical details under a single-lens microscope, recreating the procedures as they had been demonstrated by Grant. For vertebrate specimens not preserved in spirits, Edmonstone's methods were employed.

A large proportion of the specimens that he initially sent back to England for presentation were rock samples, so that on his return it was principally as a geologist that his name was made. More wonder then that he had had no real formal training as one. Applying the rules of geology, during the *Beagle* voyage Darwin derived an accurate explanation of reef formation – before even seeing a coral reef! He did so by applying Lyell's explanation of uplift and subsidence to give the first tenable account of atoll formation, providing a geological question with an organic answer; corals grow to great depths on the periphery of sinking oceanic islands. Darwin's ideas were presented piecemeal in three stages: a brief statement to the Geological Society of London in 1837, a description of reefs visited published in *Journal and Remarks* in 1839, and in much greater detail with *The Structure and Distribution of Coral Reefs* from 1842, the final, book version; 'though a small one, [it] cost me twenty months of hard work, as I had to read every work on the islands of the Pacific, and to consult many charts'. Thus, as with his other predictions, Darwin had first made sure of his facts. And again, as with his other predictions, this one also offers rather useful insights – for example, if the land continues to subside it can lead to the flat-topped seamounts that can be seen in submerged archipelagos.

Alexander Agassiz (1835–1910) should have known better having witnessed his father's 'run-ins' with Darwin over evolution and race. Coincidentally it was a visit to Edinburgh in 1876 that inspired him to choose the atoll for his gladiatorial arena, leading eventually to him spending his last 20 years travelling to every major coral reef collecting evidence in an attempt to refute Darwin and avenge his father. Although initially a reluctant antagonist, Agassiz was still driven in his quest even after Darwin's death, writing in 1902: 'Such a lot of twaddle – it's all wrong what Darwin has said, and the charts ought to have shown him that he was talking nonsense [...] At any rate I am glad that I always stuck to writing what I saw in each group and explained what I saw as best I could, without trying all the time to have an all-embracing theory. Now, however, I am ready to have my say on coral reefs and to write a connected account of coral reefs based upon what I have seen'. His alternative 'connected account' involved the formation of comparatively thin layers of coral atop shallowly submerged land, produced as the net effect of subsidence, uplift and erosion. But Agassiz strangely failed to publish his findings despite making promises for years, and seeming increasingly ready to do so from 1907 onwards, when he stated, 'I have started on my coral-reef book [...] I have made a fair beginning [...] I have worded hard', and also revealing in 1910 that he had already sketched out the book at least three times over. A few days later he was dead, leaving no sign of a manuscript. It would have mattered little: drilling technology able to reach far enough below the subsidence-generated, deep coral layers has since shown that Darwin was pretty much spot on. So convinced was Darwin by his evidence that he was even able to predict future discoveries.

It is good to know that Scotland has actively celebrated Darwin's involvement with coral reefs, albeit indirectly. The 'Darwin Mounds' are 100 km² of undersea sand that provide a habitat for cold-water, deep-sea coral reefs, 1 km under the sea and 100 nautical miles (185 km) northwest of Cape Wrath, off the northwest coast. The uniquely shaped 'teardrop' mounds were discovered in 1998 using remote sensing from the Royal Research Ship *Charles Darwin* (hence the name), now decommissioned after her 787 000 nautical miles, otherwise 32 times around the world, during 21 years of service. Because they are deep-sea reefs they are comparatively fragile, unable to cope with disturbances such as from waves, and so have been protected since 2001, especially from permanent damage by dragnet trawling which has been legislated against since 2004.

Paul Pearson followed in Darwin's footsteps, recognising his contributions to geology, and his long-term fascination with corals:

I was first influenced by Darwin as a schoolboy when a teacher lent me his battered old copy of the *Origin* [*of Species*], which caused a complete turnaround in the way I viewed the world and made me want to be become a scientist (thanks Mr Doyle, wherever you are!).

As a geology student I became interested in Darwin the man. One of my main interests at the time was what is generally known as the 'evolution' of igneous rocks. This is how (for example) a basaltic liquid often fractionates towards a more silica-rich composition along what is known as a 'line of descent' by the growth and removal (e.g. by sinking) of silica-poor crystals. I was amazed to find that Darwin had anticipated this theory in some detail. I see it as a process that is analogous in important respects to natural selection, and on researching the subject more thoroughly[2] I was delighted to discover that Darwin had based his ideas on observations that he made in the Galapagos Islands. Moreover he had worked it up at exactly the same time (spring 1838) as he was initially developing natural selection. The theory is still central to igneous geology and is one of Darwin's more important 'unsung' contributions to science.

In the early 1990s I was part of a geological expedition drilling flat-topped seamounts in the Pacific Ocean. The model for their origin was an extension of Darwin's subsidence theory for coral reefs (these were 'dead' atolls that had developed on volcanoes, as Darwin had suggested, but continued to subside after the reefs had been drowned at some point in the past). I remember the scientific crew toasting Darwin's memory when, from under over a kilometre of ocean and hundreds of metres of coral limestone, we pulled up cores of sun-baked lava flows from a long-lost Cretaceous volcanic island.[3]

More recently I took a trip to the Cape Verde islands (the first stop on the *Beagle* voyage) with my friend Chris Nicholas to investigate Darwin's geological observations there. These, he later recalled, were important in setting themes for his geological research career. We found that Darwin's 'conversion' to the gradualist geology of Charles Lyell was less immediate than has generally been thought. But what really pleases us is that when writing his autobiography as an old man, Darwin described the exact spot at which he first decided to write a book on the geology of the voyage, thereby initiating his formal career as a scientific researcher. It is a remote and spectacular place, with beautiful rock pools just as Darwin described, and as far as I know we are the only ones to have visited it with Darwin in mind.[4]

Another brilliant geologist who made contributions to biology was James Hutton, who is known chiefly for his views on geological time. Hutton was a central figure in the Edinburgh Enlightenment, and a friend of Charles's grandfather, Erasmus Darwin. It has long been known (although not well known) that Hutton articulated the concept of natural selection in an unpublished work on agriculture. In 1794 he also published the same ideas, in a rather more structured way, in the middle of an obscure three-volume treatise on philosophy. I was amazed to find this published account while researching for an article on Hutton's agricultural experiments, and so brought it to general attention through a note to *Nature*. Predictably some Creationists, thinking the worst of human nature, have accused Darwin of plagiarising Hutton. However anyone who works with ideas knows that concepts can come and go in many guises before someone produces them in the right form and provides the evidence to turn them into serious theory. There is no evidence for any direct link between Hutton and Darwin (except in geology, where there was a significant influence). There may have been some indirect link, as the selection meme filtered its way through the Edinburgh university clubs and societies, or through Darwin's family connections, but that is just speculation. Whatever is the case, Hutton's brilliance in no way undermines Darwin's stupendous contribution to human thought, just as Darwin's insights about igneous rocks do not detract from the significant theoretical advances that were made by 20th-century petrologists who took a detailed interest in the same basic questions that had intrigued Darwin a century before.

During Joseph Hooker's own maritime voyage, Darwin became something of a role model, in whose footsteps Hooker aspired to follow. So it must have been quite an honour when, upon his return in 1843, Darwin approached Hooker to take on the botanical description of many of the *Beagle* specimens. Darwin had hoped that Henslow would be able to take on this work, sharing the burden with William Hooker, but progress was scant, therefore Darwin turned to the recently landed Joseph. So started a close and trusting friendship that would last until Darwin's death.

Just how quickly Darwin came to trust Hooker is evident from a letter admitting to his earliest convictions about transmutation, written on 11 January 1844, only a few months after their first correspondence:

> I have been now ever since my return engaged in a very presumptuous work & which I know no one individual who w^d not say a very foolish one. I was so struck with distribution of Galapagos organisms &c &c & with the character of the American fossil mammifers, &c &c that I determined to collect blindly every sort of fact, which c^d bear any way on what are species. I have read heaps of agricultural & horticultural books, & have never ceased collecting facts At last gleams of light have come, & I am almost convinced (quite contrary to opinion I started with) that species are not (it is like confessing a murder) immutable.

Hooker was not from a wealthy background – and was not a celebrity when he returned to England following his voyage – so in addition to taking on Darwin's specimens his main priority was finding a steady income. His application, in 1845, to become Professor of Botany at the University of Edinburgh was rejected, and instead of accepting an equivalent chair being offered him at Glasgow University he received his father's help to join the Geological Survey of Great Britain in 1846 as a palaeobotanist, to search for fossils in the coal-beds of Wales. By the next year, and although engaged, Hooker was still keen to travel and study foreign nature, and again his father was able to help, using his position as the first director of the Royal Botanic Gardens at Kew. Hooker secured a place aboard an expedition to collect specimens in the Himalayas and India, that would run until 1851. This time, on his return, Hooker did marry his patient fiancée, Henslow's daughter, Frances Harriet.

In April 1856 Darwin invited his next-door neighbour John Lubbock (1834–1913), leading entomologist Thomas Vernon Wollaston (1822–1878), along with Huxley and Hooker, to Down House. He gave them a tour of the house and gardens before speaking to each in private about his ideas on transmutation, a highly contentious subject. All except Wollaston agreed with Darwin's ideas, with Hooker going further in trying to persuade Darwin to publish them, at least in the form of a 'Preliminary Essay'. Although Lyell didn't fully agree with those ideas, he had made the same suggestion to publish when Darwin had also confided in him, during his visit to Downe only a week before. Although very appreciative of Hooker and Lyell's support, Darwin was not in favour of their proposal, firmly believing that it would be 'quite unphilosophical to publish results without the full details which have led to such results'.

Subsequent to this espousal of the ideas within the *Origin of Species* years before it was even published, Hooker was ever compelled to defend his friend's work. Indeed, years later, at probably the most famed confrontation between Darwin's critics and his supporters, Bishop Samuel 'Soapy Sam' Wilberforce (1805–1873) traded insults with Huxley at the Oxford University Museum, on 30 June 1860, the day after Owen and Huxley had duelled once again over human origins. Legend has it that, towards the end of a lengthy speech, Wilberforce, who had been primed by Owen, asked Huxley whether it was on his grandmother or grandfather's side that he claimed to be descended from a monkey. Huxley famously quipped that he would rather be descended from a monkey than associated with a man who used his great gifts to obscure the truth. In the ensuing chaos, Robert FitzRoy, now an Admiral, holding aloft a Bible, 'implored the audience to believe God rather than man'. Writing to Darwin only a couple of days later, Hooker is clear that it was he who rescued his old friend from Wilberforce's tirade:

> The meeting was so large that they had adjourned to the Library which was crammed with between 700 & 1000 people, for all the world was there to hear Sam Oxon. Well Sam Oxon got up & spouted for half an hour with inimitable spirit uglyness & emptyness & unfairness [...] he ridiculed you badly & Huxley savagely. Huxley answered admirably & turned the tables, but he could not throw his voice over so large an assembly, nor command the audience; & he did not allude to Sam's weak points nor put the matter in a form or way that carried the audience. The battle waxed hot. Lady Brewster fainted, the excitement increased as others spoke – my blood boiled, I felt myself a dastard; now I saw my advantage. I swore to myself I would smite that Amalekite Sam hip & thigh if my heart jumped out of my mouth [...] so there I was cocked up with Sam at my right elbow, & there & then I smashed him amid rounds of aplause. I hit him in the wind at the first shot in 10 words taken from his own ugly mouth & then proceeded to demonstrate in as few more 1 that he could never have read your book & 2 that he was absolutely ignorant of the rudiments of Bot. Science [...] Sam was shut up – had not one word to say in reply & the meeting was dissolved forthwith leaving you master of the field after 4 hours battle. Huxley who had borne all the previous brunt of the battle & who never before (thank God) praised me to my face, told me it was splendid, & that he did not know before what stuff I was made of. I have been congratulated & thanked by the blackest coats & whitest stocks in Oxford (for they hate their Bishop quite [...]) & plenty of ladies too have flattered me [...]

8

TIME OF DEATH

So, to heaven's gates the lark's shrill song ascends

Darwin was quick to appreciate that the history of nature had been captured, literally, within the geological record. Much later in life, in about 1865, when writing to Frederic William Farrar (1831–1903), a master at Harrow School who was principally corresponding about language, Darwin forms a striking analogy with the geological record: 'Considering what Geology teaches us, the argument from the supposed immutability of specific types seems to me much the same as if, in a nation which had no old writings, some wise old savage was to say that his language had never changed'. This is a noticeably sarcastic response to Farrar about 'the *supposed* immutability of specific types'. From his letters, Darwin was happy enough to engage in protracted correspondence with very many people, some quite ignorant in their criticisms of his work. But, having delayed publication of the *Origin of Species* for as long as it took to be confident in his data, once convinced of the truth he seemed impatient with those whose arguments showed their ignorance of the evidence. Towards the end of his life, his exasperation was evident: 'I am tired to death with writing letters; half the fools throughout Europe write to ask me the stupidest questions'.

Their paths would cross one last time when Farrar, now Canon of Westminster and Rector of St Margaret's, resolved the issue over the agnostic Darwin being buried within the Abbey[1] by suggesting a petition for his interment which succeeded, and then acted as a pall-bearer, alongside Huxley, Hooker and Wallace, before delivering the funeral sermon. This was not a clerical coup. Rather, theirs was a relationship based on mutual admiration and respect, facilitated by the clergyman's personalised doctrine in seeking God's truth, that 'Science is itself one of the noblest forms of Theology', an approach that earned him the crown 'science's most ardent champion', second only to Huxley. Farrar may have seen harmony between science and religion, but nonetheless only four days later Harvey Goodwin, the Bishop of Carlisle, preached from the same pulpit: 'He [Darwin] observed nature with a strength of purpose, and a pertinacity, and an intensity, and an ingenuity, which has never been surpassed', and almost in the same breath hijacked Darwin's genius for the sake of the Church, with glaring circularity:

> Surely as we walk round this church and look at the monuments which it contains – memorials of great poets, and great statesmen, and great discoverers, and great commanders, and patriots, and philanthropists, memorials of the good and great in all the varied forms in which greatness and goodness happily manifest themselves in this world, which contains so much that it is neither great nor good – surely the impression made upon our minds must be that these great and good men are nothing else but the gift of God. Shakespeares and Newtons would be inconceivable in a world that was not governed by a loving father [...] if I try to devise a theory of man, and his origin, and his destiny which shall not involve belief in an Almighty and All-merciful God, it is man himself that stands as the greatest and most insuperable difficulty in the way. Great and good men seem to me to postulate the existence of an all-great and all-good God.

As we shall see later on, assuming contradictory starting points still underlies much of the science–religion divide. However, there are also debates within sciences, and within religions, so starting from the same point can also cause ructions. Within evolutionary science, these tiffs are usually temporary and are usually quietly resolved in the pages of scientific journals. But one fracas, over what is widely thought of as a scientific challenge to Darwinism, has persisted quite publicly, because it suggests an alternative, stop–start pace of evolution, so undermining the core tenet of Darwinian gradualism.

9

PUNCTUATED EQUILIBRIUM

How quick time is flying, how keen fate pursues!

Some would say that punctuated equilibrium is an embarrassment to science. Rhetoric versus dialectic. The persistent attraction of the idea is mainly thanks to the eloquence and prodigiousness of Stephen J. Gould's, the main proponent's, popular science writing. That much is widely accepted. Nonetheless, if scratched, the irritation over punctuated equilibrium sometimes, uncharacteristically, can break the surface. Gould's long-term adversary John Maynard Smith (1920–2004) wrote of him:

> Because of the excellence of his essays, he has come to be seen by non-biologists as the preeminent evolutionary theorist. In contrast, the evolutionary biologists with whom I have discussed his work tend to see him as a man whose ideas are so confused as to be hardly worth bothering with, but as one who should not be publicly criticized because he is at least on our side against the Creationists. All this would not matter, were it not that he is giving non-biologists a largely false picture of the state of evolutionary theory…

And elsewhere: 'Stephen Gould is the best writer of popular science now active […] Often he infuriates me, but I hope he will go right on writing essays like these'. This 'false picture' stems partly from widespread misquotation of a single paragraph in a Gould essay, particularly by anti-Darwinists who think that they have excavated a confession of fallacy in the fossil record from a leading scientist. Of course, they are wrong, and are prostituting the quotation out of context; Ben Goldacre, in his book *Bad Science*, terms such selective citing as 'cherry picking'. That devious banality aside, Gould's message of punctuated evolution would appear to be as challenging to a neo-Darwinist as claiming that fossils are a Divine dupery. To avoid hypocritical misrepresentation, I am compelled to reproduce the relevant extract[1] in full:

The extreme rarity of transitional forms in the fossil record persists as the trade secret of paleontology. The evolutionary trees that adorn our textbooks have data only at the tips and nodes of their branches; the rest is inference, however reasonable, not the evidence of fossils. Yet Darwin was so wedded to gradualism that he wagered his entire theory on a denial of this literal record:

The geological record is extremely imperfect and this fact will to a large extent explain why we do not find interminable varieties, connecting together all the extinct and existing forms of life by the finest graduated steps. He who rejects these views on the nature of the geological record, will rightly reject my whole theory.

Darwin's argument still persists as the favored escape of most paleontologists from the embarrassment of a record that seems to show so little of evolution directly. In exposing its cultural and methodological roots, I wish in no way to impugn the potential validity of gradualism (for all general views have similar roots). I only wish to point out that it is never 'seen' in the rocks.

Paleontologists have paid an exorbitant price for Darwin's argument. We fancy ourselves as the only true students of life's history, yet to preserve our favored account of evolution by natural selection we view our data as so bad that we never see the very process we profess to study.

For several years, Niles Eldredge of the American Museum of Natural History and I have been advocating a resolution to this uncomfortable paradox. We believe that Huxley was right in his warning [to Darwin on the eve of publishing the *Origin of Species* that 'you have loaded yourself with an unnecessary difficulty in adopting *Natura non facit saltum* so unreservedly']. The modern theory of evolution does not require gradual change. In fact, the operation of Darwinian processes should yield exactly what we see in the fossil record. It is gradualism we should reject, not Darwinism.

So Gould's piece is in actual fact an argument for evolution as evidenced by fossils. Gould and Eldredge were not really claiming an alternative to Darwinism – punctuated equilibrium is not saltation – but a suggestion for how natural selection operating at a variable rate could have produced the observed palaeontology. This wasn't because they didn't believe in natural selection as proposed by Darwin, nor in its power to produce variation over relatively short periods. In fact, rapid diversification produced by natural selection was at the very heart of their idea. What they were trying to explain away is the problem that they had with their perceived discontinuity in the fossil record, but, critically, it is a problem that neo-Darwinists do not share.

Nor, Dawkins and others tell us, did Darwin ever commit to a constant speed for evolution,[2] just a gradual speed compared with that of the saltation alternative. Sometimes the meander may even be punctuated by a stasis; in the *Origin of Species* he writes: 'Species and groups of species which are called aberrant, and which may fancifully be called living fossils, will aid us in forming a picture of the ancient forms of life'. To claim that gradualism operates at a constant rate is an unfortunate misreading of the *Origin of Species*. Additionally, and probably most unfortunate for Gould and Eldredge, the cost to veracity may have been too high in communicating their ideas to a general public, especially when assuming representation on behalf of colleagues: 'All paleontologists know that the fossil record contains precious little in the way of intermediate forms; transitions between major groups are characteristically abrupt'.[3] Such an apparently damning statement from a colleague is irritating, but only as bad as it was for Gould, being quoted out of context.

Such are the charges made against Gould and Eldredge, and especially Gould. But, of course, as venomous as it gets, the argument over punctuated equilibrium revolves around a central question common to biologists, namely how new species are produced. The star of speciation as far as Darwin was concerned is natural selection, but he lacked genetics and so wasn't able to expound all the details. Nowadays, evolutionary biology has a huge array of techniques at its disposal, and two recent studies in particular have helped considerably by contributing quantitative evidence for Darwinian mechanisms of speciation.[4] They show how species adapting to differing habitats become more different the longer they spend separated, and how that adaptation can promote speciation by modification of a single inheritable trait (e.g. camouflage).

Darwin's own understanding of speciation was founded on the inheritance of characteristics developed through Lamarckian use. These, we now know he wrongly believed, were transferred to offspring as 'gemmules' (a name suggested by Grant) which originate in the tissues and collect proportionally in the reproductive organs ready for fertilisation. But this was essentially just a revision of pangenesis dating from Ancient Greece, primarily through Democritus (*c*.460–*c*.370 BC) and Hippocrates of Cos (460–377 BC), for which we can be grateful for the names 'gene' and 'genotype'.[5] Thanks to developments during, and since, the modern evolutionary synthesis, we also now know that inheritance isn't the only source of variation. The process of natural selection acting on random mutations now shares the limelight with genetic drift. How important each is for a population depends on the number of individuals, and, of course, the varying rate of Darwin's natural selection. Brian Charlesworth has been at the forefront of the debate on punctuated equilibrium:

> I can't really remember when I first read the *Origin of Species* now, it was quite a long time ago. And, obviously, that's probably the most important single book written about biology, and nobody who's interested in evolution can deny that. It's remarkable how insightful Darwin was when you read the book, he really did think of nearly everything, nearly all the problems in evolutionary biology. Of course, he was, I think, hampered in establishing natural selection

as a sort of totally convincing theory because he didn't understand how inheritance worked, and in fact believed in blending inheritance which creates a tremendous difficulty for natural selection because natural selection, as Darwin made abundantly clear, is basically a mechanism for transforming variation within a species and differences between species.

The whole immutability of species was the problem in his day, and he decided on pangenesis and a Lamarckian type of conversion.

Well, Fisher in the introduction in the first chapter of the *Genetical Theory of Natural Selection* goes in some detail into Darwin's problems with blending inheritance, and in fact I think it was pointed out by the Professor of Engineering in Edinburgh, Fleeming Jenkin, that there was this tremendous difficulty and Darwin had overlooked it, and was rather shaken, I think. It was in a review of one of the editions of the *Origin* [*of Species*], that he hadn't appreciated that blending inheritance was as variable as it would appear, and he developed these ideas about inheritance that are quite characteristic in response to that, very much in response to that. And, of course, Fisher documents how, when the correct basis for inheritance was established, this problem simply evaporates. There is no tendency for variation to disappear. So, I think it's absolutely crucial in fact for the understanding of natural selection to have a correct theory of inheritance. It's not a matter of 'dotting the i's and crossing the t's'.

So, do geneticists feel absolutely submerged in natural selection?

Well, there are geneticists and geneticists. Guys over in the Swan Building [part of the University of Edinburgh] who can isolate mutants controlling the cell cycles of developmental pathways probably don't think too much about natural selection. But, when they do comparisons between species for their sequences of the molecules that they're interested in, and find that the amino acid sequences are strongly conserved whereas the silent sites are not, that's evidence that natural selection is actually winning out against deleterious mutations.

But as an evolutionary biologist, I should say that basically what genetics has done is two things. One, is to show that the theory of natural selection has an extremely strong theoretical basis because it is a mechanism which will work and work; moreover, one of the preoccupations of Haldane, in particular, was showing that it will work within a credible timescale. If natural selection can work, but it will take ten billion years to cause a favourable mutation to spread through a population, then we would be in bad trouble. As it happens it doesn't, and that can be shown by some fairly simple calculation. The other thing of course which genetics has done is to show there are alternative mechanisms for evolution, particularly random genetic drift which just depends on random sampling fluctuations in small populations. Now Darwin has a few hints that he was aware of such a possibility – I mean, one or two places in the *Origin* [*of Species*] – but it was never developed as a sort of very strong element in his thinking. And, of course, until you confront variation in DNA sequences which does not affect the phenotype directly, you have no real reason to be very interested in neutral variability. It's really only since the rise of molecular biology that people have taken much notice of this as a phenomenon.

Technology gives us the sources of evidence that otherwise weren't there for Darwin.

It certainly gives you a very powerful tool for looking at phylogeny. I think it's completely blowing away the more traditional morphological approach to the relationships between animal and plant groups.

You hinted at gradualism; I've brought my one prop here. It's a 1982 New Scientist [no. 1301], do you remember this? You're in with other notable names. I'd like to remind you of this article you wrote.

That was largely to annoy Stephen Gould.

You do argue that all the mechanisms are there in sufficiency, without punctuated equilibrium.

Well, without evoking special explanations, etc., punctuated equilibrium, I think it's at least a partially valid description of phenomena; it's just a question of how do you explain it? And Darwin as well had problems of sudden appearances of species in the fossil record, and so on. He did tend to explain it away due to gaps, incompleteness, and there's a lot of truth in that. But we now know of course that natural selection can be a lot stronger than Darwin ever thought of, and can produce changes in allele frequencies, in industrial melanism and things like that, in a space of time which is a mere instant, geologically.

This article you wrote almost 25 years ago. But where have we gone since? Obviously, we've lost Gould.

We've lost Gould. I don't think Gould ever had a huge impact on the mainstream evolutionary biology community is my impression. He did have quite an impact on the public, and on the palaeontological community, but even the palaeontologists, most of whom know very little about evolution at a mechanistic level, because they are trained as geologists, not as biologists, a lot of them particularly in this country were pretty hostile to Gould, but a lot of them were pretty vocal supporters as well. But I think people who study evolution, genetics, systematics and at the field ecological level really didn't have much time for Gould's idea.

So, an interesting quirky parallel?

Well, it's pretty clear Gould wanted to create a new paradigm, and become a second Darwin. He was not short of an ego or two! [...] As John Maynard Smith once said, just because a theory is orthodox doesn't mean that it's not wrong. So, I mean in many ways, I think that the consensus in evolutionary biology would be that Darwin's view that natural selection is the force shaping evolution at the phenotypic level is essentially correct, and that the more evidence we've got, the stronger it seems.

Do you think Darwin has misled geneticists in any way?

The major wrong turning in Darwin's thinking was his ideas on inheritance. In fact, he accumulated an enormous amount of empirical evidence showing pervasive variability and with a genetic basis, and using artificial selection quite correctly as a model for how selection could produce radical changes in phenotypes. I mean, it's still very amusing to produce a poster of the breeds of dogs which Jerry Coyne and I used to do in our evolutionary biology class in Chicago, to show what pure straightforward artificial selection can do. Palaeontologists looking at these blighters would put them in different families probably, and yet they're all dogs! And they know they're dogs, if they meet each other on the street.

Darwin was never frightened of recognising that incredible diversity.

What Darwin called 'the power of selection'. It's one of my favourite phrases. He recognised very clearly what a lot of people, including Fred Hoyle, completely failed to understand. That selection combines numerous characteristics together to produce a very complex ensemble of characteristics, which would never appear in a population, except because selection caused each individual trait to increase in frequency in a population. So that you end up with something with a probability of one-in-a-hundred-million of being there over a relatively short span of time, and he knew that's what happened in artificial selection.

Hoyle's work has been picked up again recently in trying to account for how such complexity could arise so quickly. That it had to come from elsewhere, that it had to come from a meteor,[6] because you can't have primordial life that is so complex.

Yeah, I've never understood how someone presumably so smart as Hoyle could fall into such a stupid fallacy.[7] And to claim that it had never been discussed by anybody else was quite ridiculous. It was discussed in the *Origin of Species* right at the beginning.

Is there any quirkiness in Darwin's work that might not be thematic to natural selection?

There was one thing in the *Origin* [*of Species*] where he discusses the resistance of different breeds of pigs to some crop or some root I think it is, I can't quite remember, which is poisonous to most breeds. As he says, the 'crackers' of Virginia are aware that, I think it was the black pig breed is immune to this, or is it the other way, is it the white one? I can't recall.[8] But, given people's great interest at present in resistance – plant resistance to herbivores and this kind of stuff – and genetic variation among herbivores, and their ability to tolerate different nasty products which plants are producing to defend themselves, I think it is a rather cute little observation.

He was, as you say, very observant, and he was also very good at synthesising information from a huge range of different fields, because he thought it was relevant to selection. And, I think that virtually everything he did after most of his books, even ones which superficially had nothing to do with natural selection, actually were motivated by his wish to apply his ideas to interpreting biological facts. So, the stuff about the motility of plants, the different forms of flowers in plants which is actually the foundation of the modern study of evolutionary biology, of sexual reproduction of plants, all flows out of his relentless wish to interpret the world.

He seemed very directed and very selective about what he was going to look at.

What I always come away with when I read Darwin is that he really thought about things. He didn't just collect piles of facts, but he collected piles of facts because he wanted to interpret them and think about how they all related to what's going on underneath, and that's what science should be about. I think, unlike some evolutionary theorists, he had tremendous respect for the data, it wasn't just 'pie in the sky'.

So I think my own view is that evolutionary biology is neo-Darwinian, and that even at the molecular level where people think all about neutral processes, this is not a contradiction to Darwinian ideas, it's extending evolutionary principles to an area which Darwin could have known nothing about [...] He didn't know about DNA sequences, which is lucky for us actually! And he didn't know about the laws of inheritance, which is also probably lucky for us, I guess; it means there's still some work to be done!

10

PHYLOGENETICS

Wee, sleekit, cowrin', tim'rous beastie

In the preceding chapters we have discussed some modern-day technologies that are helping us understand the mechanisms that are fundamental to Darwinian evolution, in a way made possible by the modern evolutionary synthesis. So, if the last century was the century of the gene, then the present one has already become the century of the genome. Even though the prize for the first genome to be sequenced in its entirety went to the bacteriophage FX174 in 1980, and the first independent organism was *Haemophilus influenzae* in 1995, it has been said that the genomics revolution started punctually in 2000, on 26th June, christened a 'day for the ages' by President Clinton. It was on this day that Francis Collins, the director of the National Human Genome Research Institute, announced the imminent completion of the first draft of the human genome mapping project. Rather less daring than his American counterpart, Prime Minister Tony Blair claimed it as 'the first great technological triumph of the 21st century'; maybe not much of a claim, only 178 days in. Nonetheless, it had been six years in the planning and had taken 10 years since its formal inception to sequence three billion nucleotides with the participation of six countries. And, although it was thought to be making fine progress, just as Wallace gave added incentive to the *Origin of Species* the human genome mapping project was somewhat hastened and assisted towards its goal from the surprise intervention of Craig Venter's Celera Genomics.

The term 'genomics' is famously attributed to a beer-fuelled evening spent discussing the best name for a new scientific journal. Supposedly the name stems from *genome* (*gen*e + chromos*ome*) and -*ics* (French -*iques*, Latin -*ica*, Greek -*ika*, denoting a topic of study). However, *chromosome* itself is an amalgam of the Greek words *khroma* (colour) and *soma* (body), so more accurately, genome itself should be an amalgam of yes, *gen*e, but more accurately, -*ome* (a mass or group, suggesting a collective or an entirety, a complete set of constituent units), as in the bi*ome*.[1] This would fit better the original proposal by Hamburg botanist Hans Winkler (1877–1945): 'I propose the expression Genom[e] for the haploid chromosome set, which, together with the pertinent protoplasm, specifies the material foundations of the species'. This definition touches on another ongoing debate in evolutionary biology, that of the species concept. In other words, how exactly does one define a species?

Darwin's own species concept was boldly fluid and dynamic, in contrast to the essentialist alternative.[2] His definition may not have proved itself useful in our classification systems, but he clearly expressed his understanding of what constitutes a species in terms of the continual process of speciation: time guides the hand that gradually teases apart formerly indistinguishable groups of organisms, creating and expanding gaps between them. We now talk about genetic distances between species, which again shows us how fundamental was his thinking, except that the gaps Darwin was dealing in were the expressed end products of those genes, the phenotypic differences in morphology between groups of organisms.

From the very start, speciation is the easel upon which Darwin draped his evolutionary canvas. On p. 35 of the legendary *Transmutation Notebook B*[3] he asked himself, 'Is the shortness of life of species in certain orders connected with gaps in the series of connections?', and then on the very next and most famous page which features his first known sketch of an evolutionary tree, he described four terminal branches labelled 'A' through 'D': 'Thus between A & B immense gap of relation. C & B the finest gradation, B & D rather greater distinction. Thus genera would be formed. — bearing relation'. By the time of writing the *Origin of Species* he had coloured his canvas with hues of gradualism and biogeography, so that in Chapter VI he writes:

> To sum up, I believe that species come to be tolerably well-defined objects, and do not at any one period present an inextricable chaos of varying and intermediate links: firstly, because new varieties are very slowly formed, for variation is a very slow process, and natural selection can do nothing until favourable variations chance to occur, and until a place in the natural polity of the country can be better filled by some modification of some one or more of its inhabitants. And such new places will depend on slow changes of climate, or on the occasional immigration of new inhabitants, and, probably, in a still more important degree, on some of the old inhabitants becoming slowly modified, with the new forms thus produced and the old ones acting and reacting on each other. So that, in any one region and at any one time, we ought only to see a few species presenting slight modifications of structure in some degree permanent; and this assuredly we do see.

Ultra-Darwinist Ernst Mayr formalised this approach as the 'biological species concept', building on the foundations laid by David Starr Jordan (1851–1931) and Moritz Wagner (1813–1887), in his *Systematics and the Origin of Species*, by defining species as 'groups of actually or potentially interbreeding natural populations, which are reproductively isolated from other such groups'. The biological species concept has certainly been the most enduring and widely adopted, and clearly it embraces Darwin's own thoughts on the problem of species' identification and their taxonomic classification.

But that's only half the story. By his concluding Chapter XIV of the *Origin of Species*, Darwin is ready to make what must have been, in an age of curation and order, an alarming prediction:

> When the views entertained in this volume on the origin of species, or when analogous views are generally admitted, we can dimly foresee that there will be a considerable revolution in natural history. Systematists will be able to pursue their labours as at present; but they will not be incessantly haunted by the shadowy doubt whether this or that form be in essence a species. This I feel sure, and I speak after experience, will be no slight relief. The endless disputes whether or not some fifty species of British brambles are true species will cease. Systematists will have only to decide (not that this will be easy) whether any form be sufficiently constant and distinct from other forms, to be capable of definition; and if definable, whether the differences be sufficiently important to deserve a specific name. This latter point will become a far more essential consideration than it is at present; for differences, however slight, between any two forms, if not blended by intermediate gradations, are looked at by most naturalists as sufficient to raise both forms to the rank of species. Hereafter we shall be compelled to acknowledge that the only distinction between species and well-marked varieties is, that the latter are known, or believed, to be connected at the present day by intermediate gradations, whereas species were formerly thus connected. Hence, without quite rejecting the consideration of the present existence of intermediate gradations between any two forms, we shall be led to weigh more carefully and to value higher the actual amount of difference between them. It is quite possible that forms now generally acknowledged to be merely varieties may hereafter be thought worthy of specific names, as with the primrose and cowslip; and in this case scientific and common language will come into accordance. In short, we shall have to treat species in the same manner as those naturalists treat genera, who admit that genera are merely artificial combinations made for convenience. This may not be a cheering prospect; but we shall at least be freed from the vain search for the undiscovered and undiscoverable essence of the term species.

This is an amazingly free outlook, a fluid concept of species, that should have removed the necessity for any alternative definitions. Instead, even knowing Darwin's species concept hasn't deterred a voracious appetite for non-consensual species definitions. Here is a list of species concepts compiled in 1997, that already needed comprehensive expansion by 2002[4]:

Agamospecies concept
Autapomorphic species concept
Biological species concept
Cladistic species concept
Cohesion species concept
Compilospecies concept
Composite species concept
Ecological species concept
Evolutionary significant unit species
   concept
Evolutionary species concept
Genealogical concordance species
   concept
Genetic species concept
Genic species concept
Genotype cluster definition species
   concept
Hennigian species concept

Internodal species concept
Least inclusive taxonomic unit
   species concept
Morphological species concept
Non-dimensional species concept
Nothospecies concept
Phenetic species concept
Phylogenetic species concept
Diagnosable version
Monophyly version
Diagnosable/monophyly version
Polythetic species concept
Recognition species concept
Reproductive competition species
   concept
Successional species concept
Taxonomic species concept

There is clearly an issue deciding the most appropriate basis upon which to define a species. This is an area that phylogenomics may be able to greatly assist. Other than having the potential to revolutionise medicine through dramatically advancing the genetic understanding of disease, and paving the way for gene therapy, genomics may also hold the key to clarifying our concept of the living packages of genetic material that we strive to differentiate as species. This would be a welcome positive application for a technology otherwise blighted by Frankensteinian associations with designer babies.[5]

Genomics may not be a qualitatively different science; the techniques are based on older ones but thanks mainly to technological innovation they're a lot more powerful. Pivotal is the ability for rapid DNA sequencing, and so being able to compare whole genomes, whereas previously phylogenetics has been based on small gene samples, even single genes. You probably know that you are almost genetically equal to a chimpanzee, but did you also know that you are 35% daffodil? And half banana. Which half I'll leave to you to decide, while we hear from Mark Blaxter who has been involved with genomics for over a decade.

I was a [David] Attenborough baby, but I did my first degree at the same time molecular biology was exploding on the scene, so this Zoology department [in Edinburgh] had people doing real molecular biology in the late 1970s, and people [here] are still working in the field. Other people you probably know like Ulrich Loening who was a molecular biologist here. He was spending half his time in his trees and half the time here. So for me, I've never been a natural historian, and I read Darwin, but one of the bits that got me most was this thing about endless forms; right at the end of *Origin of Species* there's this lovely paragraph about the entangled bank, and it depends on what edition it is, but almost the closing sentence is this thing about endless forms most beautiful and most wonderful have been, and are being, evolved.[6]

What I liked about this one is 'are being', saying that it's not just something that's happening in the past, it's something that's happening now, and that made me go back and look at my concept of what's a taxon, what's a species. What is a species anyway? And realising that nobody really knows ... so as a microbiologist doing zoology, doing zoological biodiversity, I almost want to get rid of species, but it's not that I don't believe in the usefulness of species, I think we counted, what, about 1200000 species of animals and plants on the planet, and there are 10 million? A hundred million? And, one, we're never going to invent names for them all; two, we're never going to see them, because most of the ones that are left are tiny. We've done counting all the big ones, and most of the ones that are left are tiny.

And so Darwin's most wonderful forms that 'have been, and are being, evolved', most of them are small, most of them are microfauna and microalgae and microplankton, all those sorts of things in terms of eukaryotes. And there's not enough morphology to divvy them up into species. There's not enough morphologists to do it anyway, so we're using molecular data to define taxa. And suddenly you get this, or I get this lovely thing from Darwin that we have no idea about the species status of these organisms that we're looking at – absolutely none.

Is the problem not being able to find distinct boundaries?

Realising how little we know, we can easily put them into categories. Defining taxa is easy. Defining what a species is, and whether it's got a biological validity, we don't know. We don't know if the little wiggly things that grow in a lump of moss on King's Buildings campus [of the University of Edinburgh] are the same species or a different species than in a lump of moss growing in Harvard or New Delhi. Just because there's no morphology and there's just not enough people working, so we use DNA to do that, and phylogenetics and stuff.

You can be very explicit about what your model of evolution is, and that's what is really nice. You can be very explicit that this is what I take; in this analysis I say that a species or a valid taxon is this, and you can tell another observer exactly what your boundaries are, rather than saying, 'well it feels to me', and 'on Tuesday I divided them this way, and on Wednesday...'.

So, the other quote I like from Darwin is this one [*below*] from Chapter II of *Origin of Species* [...] One of the things that struck me reading *Origin of Species*, once I became a professional scientist, was there's nothing in it about the origin of species. In fact 'species' isn't defined. And so I had to check for a definition of species, because I was starting to think, 'Well, what did Darwin mean?' [The quote is:] 'From these remarks it will be seen that I look at the term species as one arbitrarily given for the sake of convenience to a set of individuals closely resembling each other and that it is not necessarily different from the term variety which is given to less distinct and more fluctuating forms. The term variety again in comparison with individual differences is also defined arbitrarily and for mere convenience sake'.

That says to me that we have a very fluid idea of what a taxon is because we're doing a slice in time now, on a very small..., no we're not counting all the organisms never mind all the species [...] a very small subset of the organisms and the species. And so for me what this says is that we ought to expect dynamism, we ought to expect a fuzziness, and we also ought to expect more things in heaven and earth than our philosophy has dreamed of. I mean just a huge diversity, and that's what we're finding – everywhere we look there's more and more and more diversity.

It seems that his concept was even more modern than we really thought because latter-day techniques are now breaking down the taxonomic boundaries.

Yes! So I think ... my feeling is that a lot of what he says seems fuzzy, what Darwin says seems fuzzy until you actually look at it ... he was being deliberately inclusive; I think more than cautious, he was being deliberately inclusive and that's different from caution. Caution is deliberately fuzzy, deliberately avoiding the point, 'it might be or it might not be'. But Darwin actually says 'I know it isn't. I know that there's no such type'. Now, some species are very good species: cheetahs and lions are very good species and there's a lot of space between them and you can see that space in lots of ways: morphology and DNA, and phylogeny, and all the rest of it. But there are other taxa where we've got no idea, no idea.

I'm a parasitologist as well and one of the parasites I worked on was *Leishmania*, it's a kinetoplastid that causes kala-azar and is related to sleeping sickness[7] and *Trypanosome cruzi* (Chagas' disease[8]). And those organisms evolved in this clonal way, and yet there is a coherence to some of the disease phenotypes. So what you see is an organism evolving clonally, it's just a protozoan, it doesn't do sex like malaria. It's evolving clonally and being constrained by evolutionary pressures to have this morphology, because this morphology is actually adaptively successful for that parasite, even though it's bad for us. And you look and some of these lineages are as old as the hills, I mean literally as old as the hills, and there are lineages of Chagas' disease which apparently caused the same disease, which by standard molecular markers are hundreds of millions of years old, separated from each other. And yet we put them into one species group because they have some coherent morphology, i.e. when we get them we get Chagas' disease. But this is just a set of lineages which have been wandering through evolutionary states and essentially doing the same thing, just like herbivores essentially do the same thing, but doing it in wonderfully different ways. Or, wandering off for a bit and coming back. And I think that must also be true of the other tiny animals.

How do selection pressures work on those kinds of organisms?

All you have is selection on a genotype, and genotype A is fixed in this environment, and genotype B isn't, and genotype A wins and it will have babies, and it's the mutations that happen in those generations and genetic drift that get fixed, or not. But because they're asexual..., with us it may be that our great, great, great, great grandfathers, somewhere in that long line of people, we share one. It might be that you and I don't share a great,

great, great, great, grandfather until 3000 BC, it might be. But we know that because we're a sexually reproducing species we're somehow coherent. This is mixing in a fuzzy tube that we've been wandering down. But these asexual lineages are actually all really carefully defined lines through space: they never mix with each other. [...]

So this thing about 'endless forms most beautiful and most wonderful', we have half a million pounds from the UK research council to look at tardigrades and their developmental biology and evolution. And that is basically just because they are beautiful. I mean they are powerful organisms for looking at biology, but they're actually incredibly beautiful as well and I'm still struck by that and I'm still an Attenborough baby at heart. Whenever somebody comes into the lab with something new its straight under the microscope.

Do you think Darwin appreciated the beauty of nature?

Yes I think so. I'm struck by..., what's beauty? Yes. What is beauty? I don't mean..., I am not so taken by kittens or butterflies.

I think on a level [of increased understanding] Darwin definitely did. Darwin definitely did. His book on earthworms is an absolute labour of love. He sat and thought, I mean he actually got inside the mind of an earthworm, or inside the earthworm lineage, and thought about what it is this organism is doing, and set up all these tests to work out what it did, and watched what they did, and watched them behave. It was a real contribution, an absolutely stunning monograph called *The Formation of Vegetable Mould, Through the Action of Worms*. It was his last scientific monograph, the last before he died, and it is all based on experiments that he did around his house in Downe which is really lovely.

There's a thing called the 'worm stone'. Do you know the 'worm stone' in the back garden? That's really lovely. So the thing about the 'worm stone' is that it's still there and it's still slowly sinking into the ground through the action of earthworms. One of the things I like about that is that these are completely hidden organisms that we see when we dig the garden but we don't think about very much. But Darwin showed that every particle of soil in his garden and in the field out the back had gone through the guts of an earthworm some time in the last year, and that humans till the soil, but not half as much as earthworms. This, again, is a bit like evolution: the hidden force. Hidden to our eyes, this hidden force in the soil making it all work. Now, I really appreciate the idea that just because you can't see it doesn't mean it's not important. So that's why we do all this biodiversity stuff on the small things, the meiofauna, the 'wigglies', and parasites being a major driving force in the evolution of their hosts.

Are they really the best models for our understanding of evolution, given such a diversity?

Yes and no. I think model organisms have been essentially showing the unity, especially model animals have essentially been showing the unity of their animal line ... [for example] the fact that half of the genes that make humans have got homologs in flies which do essentially the same job. OK, they make a fly instead of a human, but they're being asked the same question, and they're giving the same answers just inside different contexts. So, that's been crucial in showing the unity of the biology in humans and animals.

I think model organisms that we have like the house mouse, *Drosophila*, and *C[aenorhabditis] elegans*, we chose them because they breed fast, and in that way they are actually weeds, they're contaminants and things that otherwise we would want to eradicate. They're good colonisers, yeah. And they've also often had very strange recent population histories so they're often quite closely associated with human disturbance. So, that means their recent evolutionary history in the last 10–50 000 years has actually been very strange compared to other animals and the standard organisms found on the planet. Most organisms have actually been put back by us ... like *Drosophila melanogaster* in Europe and North America is essentially a thing that came with the trade. It came with fruits brought from North Africa. So, it's a very recent invasion of these habitats, and they only survive in sort of peri-human, anthropo-whatever-the-word-is.

You go to Africa and *Drosophila melanogaster* is incredibly diverse, I mean really. From North America it is incredibly un-diverse. So the first studies that were done on *Drosophila melanogaster* wild populations, which were mostly in America, were done on a very un-diverse population. We think it was almost at the level of laboratory escapes; when they weren't laboratory escapes they were recent escapes from human habitation. So, looking at a long history of evolution and adaptation to a working environment, they are looking at a very explosive population from only a few founders. I think in that way they have been bad models for that period of evolution ... it's almost getting more and more interesting, from two million years ago until now, because that's the sort of timescale in which you expect speciation to happen. We can look at population divergence in the last 50 years and we can look at generic divergence with fossils and things much deeper than that, but really to look at speciation in action that's the time we're talking about.

I'm on record as saying that *C. elegans* is an excellent model animal but probably a bad model nematode. If you wanted *C. elegans* to represent the diversity of biology of nematodes, it doesn't. It doesn't represent that diversity, it's an instance of that diversity. If you want to look at the core things that nematodes do, and compare them to the core things that *Drosophila* does, and humans do, then it's great. And there's no way that you're going to get to the sort of level of understanding of how we work, or how any organism works, if we hadn't focused on a small set of organisms, if there hadn't been the giants in the *C. elegans* world, the Bob Horvitz's and the John Sulston's, and so on, just focusing on this organism and really making it work for them. If people hadn't done that, focusing on an organism, then we would still be in almost descriptive natural history.

Is it just a matter of time and manpower?

Yes. I think genomics are a classic case. I've been involved in genomics from 10 years ago when I started to try to persuade organisations to sequence parasite genomes, explicitly to help develop into vaccines and drug types of things, that I thought was useful. Because, once the genome data is there and it's public, anyone anywhere can work on it: you don't have to have a special sequencing lab, you just go to the Net. And so I thought it was an excellent way of promoting research on major tropical disease in tropical countries, just providing the data; and it's something the West could easily do because it doesn't take much money to sequence parasites.

So, we struggled and struggled, and struggled, to get these parasites sequenced, and discussion after discussion after discussion, and after about three or four years I realised, talking to people who were doing the sequencing of the human genome, that it was only a matter of time until all these things got sequenced, because people had invested so much time and money in the infrastructure for doing human genome sequencing, they weren't just going to put it in the skip when the human genome was done. They were going to realise when they had 'an' instance of 'a' human genome they were going to need all these other genomes to work out what they did.

And what will comparative genetics give us?

Already we now have a wonderful representative series of vertebrate genomes, there's more and more due, and another 60 or so vertebrate genomes left to be sequenced which are already on a list, and who knows? That's the next five or eight years. But already we have this dataset that allows us to say, 'This is what makes us human'. We don't know why it makes us human; these are the odd things that happened in our lineage. So, ourselves and chimps share these genetic features which tell us something about our population history and our evolutionary history, what sort of bottlenecks we went through.

Selection is efficient but it is not that efficient and so there are lots of relics, past selection events in the genome, which we can now start pulling out, now we have comparative genomes to do it. We can say, 'Well, these two genomes, we know that this bit doesn't code for a protein and it's not functional now, but they are far too similar to be similar by chance, these little bits. So what were they doing, what was it doing and why is it not doing it now?' And if you just had the one genome you'd just say that bit doesn't do anything, we'll throw it away, you know, it's junk.

It's standard Darwinian evolution, it's mutually drifting and behaving just like the modern synthesis would predict. In terms of the other comparative genomics, it's showing us where backbones came from, where our mode of development came from. Where animals came from as opposed to fungi. After years and years of speculation we now have data. That's really exciting. You see papers coming out where they have analysed 15 genomes to ask, 'What was the ancient geneset for a whole animal species?' And rather than speculation, it would be nice to be able to say if we had legs, and they can say these are the genesets we had and so this is probably the biology we had. Genomics is fun and it's exploding. We're just about to sequence another six nematode genomes!

11

STEADINESS—GREAT CURIOSITY

There learning, with his eagle eyes,
seeks science in her coy abode

Coincidental with Darwin's associations with Scotland, and more of a comment on Darwin's Celtic ancestors, the Celtic Welsh name 'Darwin' derives from 'derwen' which means 'oak', as in the tree (once lending itself to the naming of Darwen in Lancashire, after its oak-lined river), although an alternative reading of the name comes from the Old English, 10th-century 'Deorwine', consisting of 'deor' ('dear') and 'wine' ('friend'). In keeping with his family name, even Darwin's physicality was redolent of arboreal grandeur. One visitor to Down House recalled: 'In the soft spring morning about sunrise I looked out of my bedroom window and saw Darwin in his garden, inspecting his flowers. His grey head was bent to each bush as if bidding it good morning. And what a head! All that phrenologists had written was feeble compared with a look at that big head with its wonderful dome'.

At this point I'm rather pleased to discover a personal affiliation: my own surname is an Anglicisation of the Celtic Irish name 'Doire' which means 'oak grove', but that's as deep as my roots get. On the other hand, the oak was adopted as England's national tree,[1] a long-standing respect reflected by its use in language: 'as sturdy as an oak' is used to describe someone who is steadfast and honourable. This association likely stems from a historical use of the tree's wood for building purposes – for timber-framed structures and tall ships. (The Latin for such strength of character is 'robustus' which gives the oak species its scientific name, *Quercus robur*.)

One such ship, with larch and oak planking on oak frames, was the HMS *Beagle*. She had been originally built as a 10-gun brig-sloop, gaining immediate celebrity soon after her launch in 1820, as part of the coronation festivities for King George IV. Five years later, the HMS *Beagle* was modified into a barque, with four guns removed and a mizzen mast added, for her first duty to accompany the larger HMS *Adventure* on a hydrographic survey of Patagonia and Tierra del Fuego. It was to be an ill-fated mission: two years into the voyage Captain Pringle Stokes, commander of the HMS *Beagle*, committed suicide. Flag Lieutenant Robert FitzRoy, aide to Stokes' temporary replacement, steered the HMS *Beagle* home in October 1830. An extensive rebuild, including replacement of her upper deck, prepared the HMS *Beagle* for her return to Tierra del Fuego, this time captained by FitzRoy, and on which Darwin sailed in December 1831.

Only six months after their return in October 1836 the HMS *Beagle* set out once again, this time on a survey of Australia. In September 1839 the new captain, Commander John Clements Wickham, who had been a lieutenant on the previous outing, named Port Darwin in honour of his ex-shipmate. The port expanded into what is now Darwin, the capital city of Australia's Northern Territory, the name being revised in 1911 when control was transferred to the Commonwealth. This third voyage concluded in 1843. In 1845 the HMS *Beagle* was refitted and became a watch vessel to combat smuggling in the Southend Coastguard District. By 1847 the ship was moored on the River Roach, but was put ashore around 1850, following complaints by local oyster fishermen that it was blocking the fairway. In the ensuing two decades, smuggling activity in the area declined to the point that the HMS *Beagle* was eventually decommissioned as *Southend 'W.V. No. 7' at Paglesham*, and sold for scrap to Messrs Murray and Trainer (possibly T. Rainer), likely being stripped down to the waterline.

Thus, all traces of the HMS *Beagle* had mysteriously disappeared, until its anchor was salvaged in 2004 when remote sensing revealed a substantial part of its hull near Potton Island, under 5 m of estuary mud. Robert Prescott, from the University of St Andrews, who led the search, has suggested that the ship's timbers were so robust that even after half a century of extreme use they may have been recycled for local building. Indeed, some well-preserved assemblies have been found in a nearby boathouse: little wonder that the Royal Navy trusted its wooden ships, a confidence epitomised in their official march, *Heart of Oak*: 'Heart of oak are our ships, jolly tars are our men, we always are ready; Steady, boys, steady! We'll fight and we'll conquer again and again'.

Alongside all his many other fascinations, Darwin was also authentically interested in things oaky; in 1880 he wrote to Sir James Paget (dubiously immortalised by association with complaints of the more sensitive areas of the body: Paget's disease of the nipple, Paget's disease of the vulva and Paget's disease of the penis[2]): 'I am delighted that you have drawn attention to galls. They have always seemed to me profoundly interesting', as well as making a number of references to them in his major works. This is not surprising; oaks and their ailments would have played an important part of his every day. Other than a fine single evergreen oak standing proud in the orchard at Down House, his 'Thinking Path', the Sand-walk, is also bordered by an avenue featuring oaks, as recounted by Francis Darwin in *The Life and Letters*…

> …he went on for his constitutional—either round the 'Sand-walk', or outside his own grounds in the immediate neighbourhood of the house. The 'Sand-walk' was a narrow strip of land 1½ acres in extent, with a gravel-walk round it. On one side of it was a broad old shaw[3] with fair-sized oaks in it, which made a sheltered shady walk…

A modern study has since filled in some detail about the Sand-walk trees: 'The upper storey consists predominantly of lime (*Tilia × europaea*), with a mixture of species between oak (*Quercus robur*), sweet chestnut (*Castanea sativa*), ash (*Fraxinus excelsior*) and to a lesser extent sycamore (*Acer pseudoplatanus*) and beech (*Fagus sylvatica*)'. So Darwin would have been close to the oak in both name and nature on every living day at Downe, walking beneath them, until his day of interment, beneath the acorns that emblazon the finials of Westminster Abbey.

As oaks, Darwins ought to be reliable, consistent types, solidly founded and steadfast in mission. Ironically, Francis Darwin remembered that it was with the help of oak that Darwin used to steady himself when fatigued: 'Indoors he sometimes used an oak stick like a little alpenstock, and this was a sign that he felt giddiness'. If not strong in body, Charles Darwin certainly thought himself strong in mind, offering 'Steadiness—great curiosity about facts and their meaning' in answer to a questionnaire asking for 'Strongly marked mental peculiarities, bearing on scientific success'. It was Darwin's observational powers that no doubt bridged that gap between curiosity and meaning. Martyn Murray provides an excellent example of them in action:

> As a postdoc in Cambridge in the 1980s, I was surprised at the number of field biologists who revered Darwin. Put simply, he was their hero, the scientist they most desired to emulate. I attribute this to the fact that Darwin was above all a field biologist himself, for he combined acute observation of the natural world with careful field[4] experiments. Take one example that can be found in the *Origin* [*of Species*] :

> "In Staffordshire, there was a large and extremely barren heath. Several hundred acres of this had been enclosed 25 years previously and planted with Scotch fir. The change in the native vegetation of the planted part of the heath was most remarkable … 12 species of flowering plant flourished in the plantation, which could not be found on the heath. The effect on the insects must have been still greater, for 6 insectivorous birds were very common in the plantations, which were not seen on the heath; and the heath was frequented by 2 or 3 distinct insectivorous birds.

Here we see how potent has been the effect of the introduction of a single tree...

...then I went to several points of view, whence I could examine hundreds of acres of the unenclosed heath, and literally I could not see a single Scotch fir ... but on looking closely between the stems of heath, I found a multitude of seedlings and little trees, which had been perpetually browsed down by cattle."

This is a surprisingly modern and fresh conclusion; it is also the first ecological description of a grazing enclosure.[5] As a field biologist, Darwin's thinking is modern, accessible and frequently right.

Even more than the answers he provides it is the questions that Darwin asks which make him so appealing to contemporary biologists. I believe he wrote somewhere towards the end of his life, that 'Looking back I think it was more difficult to ask the right question than to answer it'. His questions arose out of a combination of an inquiring mind tempered by the discipline of acute observation. From his lifetime of work devoted to evolutionary biology, I would guess that he never lost his wonder of the natural world.

Once you know what to look out for, it's possible to recognise patterns of effect in nature. Darwin was an innovator, a pioneer, and his ability to observe becomes even more amazing when you try his methods yourself. Back in his garden at Down House, he would make comparisons between different kinds of plants to test the applicability of his findings across a wider geography. That is, he was able to validly scale up from the minute to the massive in order to test his general ideas. He had already realised on Sedgwick's 1831 geological tour that 'science consists in grouping facts so that general laws or conclusions may be drawn from them'. Mighty oaks from little acorns grow, you might say. Randal Keynes has been intricately and intrinsically involved with Darwin's story and Down House, and is often left marvelling at his ancestor's comprehension of the natural world:

Yes, indeed. This I think is, well it's a very clear example of one of the qualities of Darwin that I find most remarkable, and that is his ability to look beyond what's obvious and familiar to remarkable things that lie just beyond, and can only be seen if you make an effort of imagination. On the voyage of the *Beagle* he has had the sort of life-changing experience of walking into a tropical forest in Brazil, and seeing the varieties and richness of life all around him. There are wonderful accounts that he left in his journal of researches of the voyage of that experience. When I was looking through some zoological notes that he wrote every evening in the cabin about the day's 'finds' while he was on the Falklands, some two or three years later, I discovered an account he gave of a lump of seaweed. Falklands has kelp forests on its shores, on its rocky shores, like the ones off the coast of Scotland, and elsewhere. These huge growths of kelp with these different shells and Crustacea, all sorts of different kinds, and worms and fish that lurk in them. And when Darwin on the shore saw this heap of seaweed ... I think I would just have thought of it as just a huge heap of rather smelly, disgusting matter. And Darwin writes that ... Darwin sees this lump of seaweed as it would be when the tide is in, and all the seaweed is floating in the water, and all the creatures in it were attached to the seaweed, or whatever. And he says, he sees that from the length of the seaweed, that there are huge lengths, that the kelp must grow to a great depth. 'The quantity may well be imagined', he writes, and he then goes on:

"I can only compare these great forests to terrestrial ones, the most teeming part of the tropics; yet if the latter in any country were to be destroyed, I do not believe nearly the same number of animals would perish in them as has happened in the case of kelp. (I refer to numbers of individuals as well as kinds). All the fishing quadrupeds and birds (and man) haunt the beds, attracted by the infinite number of small fish which live among the leaves. On shaking the great entangled roots, it is curious to see the heap of fish, shells, crabs, sea eggs, cuttlefish, star fish, Planariae, Nereidae, which fall out. This latter tribe I have much neglected. Among the Gasteropoda, *Pleurobranchus* is common: but *Trochus* and patelliform shells abound on all the leaves. One single plant form is an immense and most interesting menagerie. If this *Fucus* were to cease living, with it would go many: the seals, cormorants and certainly the small fish, and then sooner or later the Fuegian man must follow. The greater number of the invertebrates would likewise perish, but how many it is hard to conjecture."

And that is an extraordinary understanding from just a heap of seaweed on a shore!

12

EVOLUTIONARY COSMOLOGY

Ilk star, gae hide thy twinkling ray

Humans, real and imagined, have often looked to the stars for answers to larger-than-life questions: from Ptolemy to Hoyle, and Moon-Watcher (*2001: A Space Odyssey* by Arthur C. Clarke) to Dr Eleanor Ann Arroway (*Contact* by Carl Sagan). While we can be confident of our earthly explanations, outer space is where our logic breaks down, or when theoretical physics is forced into fantastical realms. Curvedness of the space–time continuum, the Poincaré homology sphere and the Picard horn are all large-scale properties of the universe that the Large Hadron Collider (LHC), inaugurated in 2008, hopes to clarify. Mining the very fundament of the universe will reveal the basic relations of matter, and how superstructures, including the universe, were formed. But, like looking at a pudding long after the cook is gone, understanding how the firmament appeared is somewhat different to *why* it appeared (as addressed by Aristotle's purposeful Causes). The LHC hopes to provide us with some of those reasons, partly by understanding the fabric of that superstructure, not least what can be considered to be the glue holding it all together: the celebrated 'God particle', or Higgs boson, named after its prediction by Peter Higgs,[1] a theoretical physicist at the University of Edinburgh.

Another dimension to the universe is its temporal dynamics, the turnover of inorganic, galactic matter. Evolutionary cosmology considers this renewal process to be akin to an organic process, a natural process. A Darwinian process. And, as Darwinian evolutionary biology makes no comment on abiogenesis, this new evolutionary cosmology does not attempt to explain the big question of why there is anything, rather than nothing. Notwithstanding such limits to our understanding, could the universe really conform to the same gradualism seen here on earth? Even before his childhood's end, Richard L. Gregory had perceived the existence of these parallels:

I was brought up with evolution of the stars, as my father was an astronomer. As a boy in the 1930s, I would read [the co-founders of British cosmology] Eddington and Jeans with avidity; but although Darwin was quite often discussed, natural selection was at that time controversial and generally viewed with suspicion. The concept of design by random events, with successes and failures writing the future, was hardly appreciated, certainly not by me. Natural selection is sometimes described as mindless and lacking intelligence – but it seems to me now that the Darwinian processes are intelligent – super-intelligent – producing answers science can hardly formulate, let alone fully understand.

When Darwinian evolution is claimed to solve practically all problems of the universe, one has to ask: How did stars come into being? Darwin himself realised that his biological theory does not extend to the inorganic world, so regretfully leaves problems of Creation and development of lifeless matter an inscrutable mystery. Martin Rees tries to bridge this gap, by thinking of something akin to organic evolution for the universe itself – successive Creations and destructions gradually evolving the natural laws and matter. This is a wonderful idea. It remains to be seen whether Darwin's great insight for biology extends to the universe itself.

Incidentally, my father (born in Christchurch, near Poole) when he was a boy knew Alfred Russel Wallace, playing in the grand old man's garden. This is a tenuous link with Darwin but I treasure it greatly.

'[S]uccessive Creations and destructions gradually evolving the natural laws and matter' makes reference to Alan Guth's and Andrei Linde's ideas on a multiple universe, or multiverse, itself built upon the idea of a bubble universe. For a good analogy of the bubble universe model, think of those geothermal mudpots in places like Yellowstone, Iceland and New Zealand. Air pockets rise to the surface creating thin-skinned bubbles. Most pop, but a few bubbles survive longer, by getting stretched to implausible extents, before too finally popping.

Lee Smolin's evolutionary cosmology then posits an evolutionary mechanism underlying universe survival. If energy fluctuations in the parental 'quantum foam', the vacuum precursor of universes and the mud in our analogy, exceed a certain threshold, then an expanding, persistent bubble universe forms, otherwise a small, temporary universe blips and dies in a single heartbeat of the eternal space–time continuum. Bubble universes like our own that do survive form matter and galactic structures, to a certain extent held together by the Higgs boson, and can even propagate their own bubble children through the collapse of black holes. It follows that the more black holes a universe contains, the more offspring it can spawn, and the longer it will persist. So we have variation and a selective mechanism, the prerequisites for Darwinian evolution, and so Smolin's cosmological natural selection theory of fecund universes.

However, because each bubble arises from a fluctuating energy source, each descendant universe within the multiverse is likely to exhibit differing parameters, those physical constants and laws that we hear are so critical in their range of values to allow the existence of life: a fine-tuned universe. Statistically, there will be mostly bubbles with no life, but many fewer ought to have life like ours, and different from ours, perhaps so complex to be beyond imagination. Perhaps with a greater intelligence. Let us hope that heightened intelligence is a part of human futures, perhaps so that humans can have any future prospects at all. Science fiction often predicts a dystopian future of tyranny and degradation, runaway technologies, clones and post-apocalyptic mutants. If anthropogenic calamities, like that forecast for global warming, can be averted and avoided, then perhaps science can provide a more optimistic outlook.

The popular response to global warming has been for considered mitigation of nature-impacting activities, reducing carbon footprints and emissions, recycling and renewable energy. However, there is a seemingly paradoxical view that human progress, in the form of technology, will actually emphasise our dependencies on our environment, rather than separating us from natural processes, as is typically believed. Stuart Blackman argues that such environmentally deterministic thinking, ironically, yet inevitably, will lead to our history being determined by that very environment:

> An unfounded sense of crisis dominates public discussion of environmental issues, and shrill demands for urgent action to mitigate climate change thrive at the expense of genuine, illuminating, nuanced debate about how to make the best of an uncertain future. The consensus view that we mitigate against anthropogenic climate change has important implications for both future human history and our future evolution. The story of human history to date has been one of distancing ourselves from nature. After all, it's our civilisation – our development – that has served increasingly to buffer us from the elements. And yet an emphasis on mitigation would likely serve to reverse that trend. Those who talk in terms of preventing climate change also tend to see development as the problem. The result of that way of thinking is that the human race would be left in a position where we are more vulnerable to whatever Mother Nature has to throw at us – and you can be sure she has plenty to throw, whether or not our industrial emissions are influencing the climate. In this way, environmentalism is a self-fulfilling prophecy. By bringing us closer to nature, it exposes us to environmental dangers and potentially makes natural selection an important driving force once again.

Hoping that global catastrophes are avoidable, the future of humanity remains unknown to the point that future humans could be unrecognisable. Not only will the continual selection pressures of evolution ensure some adaptive changes over time, but an accelerated form of evolution is predicted to run in parallel. This is not punctuated equilibrium, but the bioethical application of reprogenetic technology. Martin Rees has great confidence in our potential, but not necessarily in our present, human form:

Most educated people are aware that we're the outcome of nearly four billion years of Darwinian selection. But many tend to think of humans as somehow the culmination of this process. Astrophysics tells us, however, that our sun is less than halfway through its life span. It will not be humans who watch the sun's demise six billion years from now: any creatures that exist then will be as different from us as we are from bacteria or amoebae. There's more time ahead, for future evolutionary change, than the entire emergence of our biosphere has needed. Moreover, evolution is now occurring not on the traditional timescale of natural selection, but at the far more rapid rate allowed by modern genetics, intelligently applied. And post-human life has abundant time to spread through the galaxy and beyond. Even if intelligent life is now unique to earth, it could nonetheless become a significant feature of the cosmos. Our tiny planet could then be cosmically important as the 'green shoot' that foliated into a living cosmos.

I believe we are part of some marvellous evolutionary process which still has a long way to go beyond the human stage, here on earth and far beyond... Extra-terrestrial life will use genetic engineering to quickly modify themselves into new post-human species better adapted to an alien habitat.[2]

13

CREATIONISM

We bless Thee, God of Nature wide

We stereotypically imagine evangelical bigots and six-day literal Young Earth Creationists to all be of the fire and brimstone type. This was dramatically encapsulated in the character of the Reverend Jeremiah Brown in the play and film depiction of the Scopes Monkey trial, *Inherit the Wind*,[1] who preaches a creed based on the fear of God and the punishment of sinners. He is particularly zealous during one scene set at a prayer meeting, when putting his own spin on Genesis 2:7:

> BROWN: And he blessed them all. But on the morning of the sixth
>   day, the Lord rose, and His eye was dark, and a scowl lay
>   across His face. (*Shouts.*) Why? Why was the Lord troubled?
> ALL: Why? Tell us why! Tell us the troubles of the Lord!
> BROWN: (*dropping his voice almost to a whisper*): He looked about
>   Him, did the Lord, at all His handiwork, bowed down before
>   Him. And He said, 'It is not good, it is not enough, it is not
>   finished. I ... shall ... make   ... Me ... a ... Man!'
>   (*The crowd bursts out into an orgy of hosannahs and waving arms.*)
> ALL: Glory! Hosannah! Bless the Lord who created us!

Such vehemence would not seem to make any room for Darwinian science which, as we have learnt, is deeply rooted in the Greek empiricism reinvigorated by Hume's Enlightenment. Stephen Fry[2] sees an incompatibility between the two methodological frameworks of naturalism and supernaturalism, as informed by scripture:

> I've always been extremely uncomfortable with the idea in any society that
> belief is based on revealed truth, that is to say, on a text, like a Bible or
> Qu'ran. It seems to me that the greatness of our culture, for all its incredible
> faults, is that we have grown up on the Greek ideal of discovering the truth.
> Discovering by looking around us, by empirical experiment, by the combination

of the experience of generations of ancestors who have contributed to our sum knowledge of the way the world works.

The idea of revealed truth is certainly driving the 'black and white thinking' portrayed in *Inherit the Wind*. Jason Lisle, an astrophysicist with Answers in Genesis-U.S., a non-profit Christian apologetics ministry with a particular focus on Young Earth Creationism and a literal interpretation of the Book of Genesis, suggests that there is a preconception that underlies this dichotomy:

> We all have a way of looking at the world, we have sort of 'mental glasses' [...] that affects how you see things. We can ultimately look at the world through what we call 'evolutionized glasses', and I don't necessarily mean that you believe in evolution, but maybe your way of thinking about things is based on man's opinion, because [...] it's really about God's word versus man's opinion. Or, we can ultimately look at the evidence through 'Biblical glasses' [...] God's word as truth.[3]

Echoes of Admiral FitzRoy at Oxford in 1860, perhaps. But, as we shall see later in this chapter, even from within the same ministry, contemporary Creationist involvement with Darwin doesn't actually accord to such an obvious for-or-against dichotomy. Although, partly thanks to sound bites like the one above, one might be forgiven for thinking that it does. One might be excused for thinking that this is the case, for example, for the most populous denomination in Scotland.

The 2001 census gave the largest religious group in Scotland as the Church of Scotland, with 42% of the population (a total of 65% of the people being Christian). *Life and Work*, the Church of Scotland's magazine, for March 2009 gave thanks to 'Mr Darwin', because, 'As the birth of Charles Darwin, author of "Origin of the Species", is celebrated, Christians should not see this as a threat [...] Darwin's work drives us back to acknowledge the awesome majesty of God. Darwin reminds us [...] that our faith in God has to be an all or nothing affair. The ordinary is extraordinary and we look not for God in the occasional miraculous intervention for the whole of creation must live and move and have its being in Him'.

This fundamentalist response to Darwin's bicentenary is in keeping with the Church of Scotland's historically puritanical outlook, and consistent with their Confession of Faith laid down in 1647.[4] Richard Holloway, former Bishop of Edinburgh in the Scottish Episcopal Church, well-known radical, and supporter of liberal causes, even sees such unyieldingness as an unlikely source of the church's longevity:

> When Thomas Kuhn published *The Structure of Scientific Revolutions* in 1970 he provided historians of ideas with a useful heuristic instrument, paradigm theory, for the study of culture. Of course, the paradigm idea of all ideas was provided by Charles Darwin in his discovery of evolution. Paradoxically, given the enormous anxiety the discovery provoked, the idea of evolution has been

particularly useful to the historians of religion. It has provided them with an invaluable insight into the organic and developmental nature of religion, and its genius for adapting to new ideas and shifts in culture. Darwin, who had enormous sympathy for the difficulties religious believers faced in adapting to new discoveries, would have been delighted at the way religious expositors have made use of his discovery in helping such conservative social forms adapt to changing circumstances; though he might have raised a quizzical eyebrow at the claim that it was the Holy Spirit who was the prompter of cultural evolution. But an enormous paradox remains. How do we account for the fact that the most belligerently flourishing species of religion on earth today are those that make a virtue of their absolute refusal to adapt to changing circumstances? Why is it that in this realm alone dinosaurs still survive and stride menacingly across the earth?

While some churches may be accused of being prehistoric, the Presbyterian church in modern Britain has made some concession to change. Although not an officially endorsed line, the Church of England website recently published a posthumous apology to Darwin: '200 years from your birth, the Church of England owes you an apology for misunderstanding you and, by getting our first reaction wrong, encouraging others to misunderstand you still. We try to practice the old virtues of "faith seeking understanding" and hope that makes some amends'.

In contrast, one would expect Creationists to be the most ideologically removed from evolution, and unable to make any concession to Darwinism. But the reality, in Scotland at least, shows a convolution: a 2009 survey by the public theology think tank Theos showed 15% for Creationism, another 15% for Intelligent Design, a further 30% who think that 'evolution is part of God's design', and 34% who think that 'evolution removes the need for God' – suggesting that personal beliefs draw differentially on Creationism and evolution. So, what is the real relationship between Creationism and Darwinism? Ken Ham, the President/CEO and founder of Answers in Genesis-U.S., one of the most prominent organisations disseminating Creationism today, seemed to be the right person to ask. He made a visit to Edinburgh in September 2007:

In May 2007 there was a Gallup poll which said, of people in the USA, people believing in evolution was 49%, against 48%; and in 63% of those people who are against evolution, it is because they have a faith, and 14% because of insufficient evidence, and 46% of those people believe that God created man. So it sounds like Creationism is a popular view, either stable, or growing.

I'd say a number of things. First of all, when you have those polls, actually there have been a number of different polls done, they come out with different percentages, but they're sort of that order of magnitude. But one of the things that you'll notice about those polls is that they're not very specific in defining what they mean by Creation, and defining exactly what people believe. [For example] some of the people who would say they believe in Creation might say they still believe in Darwinian evolution. Some of them might be literal six-day, Young Earth Creationists, some of them could be 'millions of years' Creationists, who don't believe in Darwinian evolution, and so it goes on. So, I'm always somewhat sceptical of those polls, as to what they really mean, ultimately. But, at the same time, I think there are a number of indicators that the Biblical Creation movement is having an impact in America, and I think around the world, and I could give you what I think those indicators are. For instance, you know some of the museums in America, the Chicago Field Museum of Natural History, the Natural History Museum in New York. Some of the museums will have meetings about how to bolster their teaching in evolution, because what they're finding is that there are more and more people coming into the museums and questioning their exhibits. And I believe that's a direct influence from the Creation movement. For instance, just this past week, a group of Park Rangers, I don't know if you read that story, are coming to the Creation Museum because they're getting more and more people who are then asking questions. So they want to know what's going on. It's interesting to note that at some of the science conferences now, and some of the big ones in America, the topic of 'how to deal with Creationists' is now coming up as one of the agenda items. So, there are a lot of these sorts of things going on. And, you know, there's the Intelligent Design group too. That's a whole different group.

'Dealing with Creationists', that's just a phrase. But there always seems to be this kind of antagonism against Creationism, as a belief format. It kind of forces people into the apologetic stance. For Creationism, and Young Earth science in particular, it always seems that they have to be working so hard to try and find support, particularly evidential support, scientific support, to garner the scientific understanding that it seeks.

Yeah, even within the Young Earth Creationist movement there are a lot of different approaches to things. One of our approaches is, you know we're not evidential-ist in our approach, we're pre-supposition-alist in our approach. That doesn't mean we're not interested in evidences and so on, but we certainly, as a movement, have a sense that we are coming against the establishment. We understand that. We also know that the majority of scientists out there certainly won't agree with us. And even the majority of Christian scientists, or should I say church-going scientists, wouldn't necessarily agree with us.

And Christian scientists take a particular stance, for example Hovind's school of thought?

Well, some of those are really looking at heavy evidences to supposedly prove the Bible is true, whereas what we're saying as an organisation, we're pre-supposition-alists and we're saying we start with the Bible. We understand our starting point and we try to educate people to understand that. For instance someone like Richard Dawkins, he has a starting point too, his starting point is that there's no God, and the Bible is not relevant. And those who don't start with the Bible have a starting point in autonomous human reason. And so we try to distinguish very much between what we call 'Origin science' or historical science, and operational science, because

I think that's a big problem. I think it's a problem in Christian circles as well as non-Christian, because there are Christians out there who say, 'We believe the Bible, we don't believe in science'. And we say you can't say that, that's wrong! There are others in the non-Christian world, the evolutionists, who will say, 'Well, science has proven that you can't trust the Bible'; you know the whole lateral science versus the Bible. And you know, we tell them, 'Hey, we have the same science as you when it comes to operational science'.

In your presentation, that is now accessible via the internet, you use a very scientific basis; I mean there's lots of genetics in there.

Well, we do. What we're doing is saying, 'We can look at observational science and show how observational science confirms our starting point'; doesn't prove it, but confirms our starting point. That's what we're really saying.

You're in a wonderful place for, well you brought it up, 'posteriori' and 'a priori' understanding: David Hume's statue is just up the road. He called them relations of ideas and matters of fact. And I think that comes back to Dawkins, that there is supposed to be no 'a priori' assumption. He would argue that he doesn't have an atheist viewpoint to start with, it's just that his evidence and understanding of the science has led him to that viewpoint because he finds no evidence for a deity.

Yeah, but what evidence is he prepared to accept? And, I would say that what he is stating there is simply false, and he doesn't believe in the deity, and that's why he has the particular belief that he has. I mean, if you think about it from the perspective of an atheist, you have to ask him, 'What evidence are you prepared to accept?' And, how does he know that that's the right evidence anyway?

I guess he's commenting that it needs to be proven through scientific method and peer-reviewed science. Yes, it has fallibility, but it's not as bad as resorting to mysticism.

I'm going to tell him that the same is true of his evolutionary beliefs. Because his evolutionary beliefs involve the past, when he wasn't there. What about the law of abiogenesis? Inanimate matter producing life. How could that be?

Well, I mean, I don't think anyone would be really comfortable commenting on that as an evolutionary biologist. I think the realm of evolution is supposed to be after life had occurred. I agree it's part of the whole story,…

It is part of the whole story, and he can't ignore it…

…but that's the scientific truth. Dawkins' specialism is within one field.

You see to me that's an excuse, they always make that excuse. They say evolution is modification by descent, and doesn't deal with the origin of life. And I say, that's nonsense because if you haven't even got a mechanism to show how matter could produce a code system and information systems, how can you say that there is a mechanism now to continue to produce information for new characteristics and so on? And it really is a part of the whole system, and they ignore it because they don't have any evidence for that. And then what they are looking at is of course just observation regarding natural selection, speciation, and so on.

You mentioned one of your catchphrases, 'Were you there?' You tell children to challenge their science teachers. But your critics say that, if it's a valid challenge to evolution, then it's an equally valid challenge to Creationism. You say, God was there, and that's the answer.

That's right, that's exactly right. And that is the answer. What we're saying is, look, we admit that we're starting from the Bible, and we admit that we're saying the Bible is the word of God and He's the one who has always been there and the one who knows everything and He's revealed to us the true history of the world. So, that is our authority, that's true.

I think the majority of Christianity would say that there is an element of interpretation in the Bible and yet you do stand beside it, saying it's the literal word of God?

Well, we always say, there are two different ways of saying it. When someone says do you take the Bible literally? I say we take it 'liter-arily'. What do we mean by 'liter-arily'? Well, what we call a grammatical, historical, interpretive method. Of course, you're always interpreting, that's true. But the grammatical, historical, interpretive method means you're taking it naturally. That's what we mean by 'liter-arily'. In other words, is there poetry in the Bible? Absolutely, the psalms are poetic. That doesn't mean there is no truth in them, but they're poetic. When it's history, it's history; when it's poetry, it's poetry, when it's prophecy, well you know what that means. We can say Genesis is written as history. It's typical, historical narrative. It has the typical form, written as historical narrative, it's always been looked at as history. In fact the New Testament writers treated it as history, and so we take it 'as written'. And to the best of our ability, the grammatical, historical, interpretive method is letting the word of God speak to us.

Are you actually confident in your ability to tell the difference, between prose and poetry, and descriptive, accurate and representative narrative?

Well, I think there are enough experts around who can do that, and we have our own particular set of experts, Hebrew scholars and others, who help us in making sure they keep us on what we believe is the right track.

So I've been reading my Genesis this week, in preparation for seeing you, and am intrigued by a statement, very early on, [God saw all that He had made, and it was]'very good'. And I wondered, as part of that process of understanding the language, what does 'very good' mean?

Well, ... any word has different meanings depending on context, as you know. We could say 'That was a good meal', it doesn't mean it was perfect, right? But, in the New Testament someone came to Jesus and said 'Good Master' and Jesus said, 'Why do you call me "good"? There is only one good, that is God'. Obviously there's a special definition of good that is reserved for God ... as well as looking at the meaning of the word 'good', there it's in the Greek of course, and in the Old Testament they were looking at the Hebrew, it can have a number of senses. One sense is sense of perfection, and it has other senses too, but there's a special sense that's reserved for God ... when you look at the attributes of God, that He's all merciful, that He's all loving, that He's without sin, and so on. So when God is using the word 'good' or 'very good' I would say that that has to be consistent with his attributes, so it would have to be perfect, and pure, and without corruption, and so on. And that is a valid understanding of that particular word. And it's the New Testament that helps us to understand, to cross-reference, that particular statement.

For someone like Dawkins, there's just no communication there because it always has to be evidence based.

I think he's not prepared to accept that he has a starting point. He says that he works from evidence, and you're not allowed to work from the Bible. And if I was in discussion with him, I would have to force him to admit that he has a particular starting point. If he's not prepared to admit it, then I'd say he's inconsistent. I'd say that he's also inconsistent in that, if he's going to talk rationally and logically, we all accept the laws of logic and natural law, and the uniformity of nature, and so on, and I'd say it's illogical to do so if you don't believe in God. And we can talk about those sort of things as well. But I think really the reason Richard Dawkins doesn't want to discuss things with Creationists, and others, is that publicly, they've already stated they don't want to give Creationists a platform, so that people can hear their message. I would say that Darwin ... in a sense, you could say that they are using it as you would use circumstantial evidence in a murder-mystery. In other words, you can use circumstantial evidence, such as natural selection, such as speciation, and so on, and most of what Dawkins uses in regard to Darwin, of course we agree with. Because we agree with natural selection, we agree with speciation, we agree that mutations occur and so on. Where we disagree, is that there is a mechanism to produce new information to add into the genome.

You mentioned that you can accept certain elements of Darwin's observations. He was wonderfully observant and quite the humanist as well, I understand. And, of course, he came from a theological background. He is always set up as the anti-theologian, but what does Darwin mean to you?

What does Darwin mean to me? Well, I'd say that Darwin basically tried to come up with an explanation of life without God. We have to admit that Darwinian thinking is a dominant paradigm in the world today. It certainly influenced the world in a big way and it certainly is very well known all around the world.

Is it a good thing for the world?

I would say that what Darwinian evolution has done has been one of the instruments in undoing, for instance, the work of the Reformation. Darwinian evolution, here's the interesting thing, I would say that in essence Darwinian evolution is not the real issue. Because of the acceptance of atheistic, Darwinian ideas, I believe that we've seen a collapse in Christian morality in our Western world. Because, if your generations are taught that you are a result of natural processes, and that there is no God involved, there aren't implications. It's not that Darwinian evolution is the cause of that, but the more generations that have been taught Darwinian evolution as fact and that this is an explanation for life without God, then the more that have rejected the Bible. So it leads to consequences in regard to your morality.

There is a lot of good that has been done in the name of God. There is fantastic work all around. A lot of atheists would feel insulted that they would be called amoralists if they don't believe?

No, because we're not saying that. What we're saying is, if someone becomes consistent with a naturalistic view of origins, then who does decide your morality? Who does decide right and wrong? Who does decide good and bad? The point is that morality is relative, it's all subjective, and that's the point that we're making. You see, some misunderstand us. They think that we're blaming evolution for abortion, or gay marriage, or whatever. We're not blaming evolution. What we're saying is that indirectly there's blame – in that the teaching of evolution to generations as fact, that teaching Darwinian evolution, 'molecules to man' evolution, naturalistic evolution, has caused many to say therefore that the Bible can't be true. Therefore, that leads to a basis for saying well, who decides on morality? Where you come from does affect your whole life view. I mean Dawkins himself, when he was interviewed for a magazine in America, he was asked the question, 'Well, some people say that from what you teach, that life is meaningless, and hopeless', and his answer was basically, 'Well, that's tough!' He admits that that is true.

So these people, Dawkins and Darwin, are they misled?

I would say that the issue is that the heart of man is deceitful, and above all things desperately wicked. That we're born as sinners, and we're rebelling against God, and we don't want the true God. And that's what the issue is.

So are they misled?

Are they misled? Well, the Bible also says that they're without excuse. Because they know there is a God, it's written in our hearts and it's obvious from the evidence.

They're blind?

So, they're certainly blinded to it. But from a biblical perspective, it also says that blindness is 'willing blindness'. The 2nd Peter 3, they are willingly ignorant of Creation and the flood ['2 Peter 3:5 – For this they willingly are ignorant of, that by the word of God the heavens were of old, and the earth standing out of the water and in the water']. In other words, you have to override the obvious. You have to override what's written in your heart. You have to override what's obvious and all around us.

And the consequences of not accepting God? Dawkins: where's he ending up?

Where's he ending up? I'd have to put it in different words. In this way, in that where people believe in God, but from a biblical perspective, the Bible says, 'unless a man is born again, he cannot enter the Kingdom of God' [John 3:33]. 'That if you confess with your mouth, "Jesus is Lord", and believe in your heart that God raised him from the dead, you will be saved' [Romans 10:9]. So you can believe in God, but the Bible even says the devils believe in God too!

So, would you have him as the Chaucerian devil's chaplain?[5]

Well, what I would say is, that because of their own ignorance, that they are not prepared to put their faith and trust in the Lord Jesus, that they'll be eternally separated from God. Dawkins is wrong, he thinks that when we die, that's the end of it. I'd say no, that's not.

Talking about the scientific message that [Dawkins] believes in so strongly: what scientific evidence is acceptable? Science has done amazing things over time: space travel, for example. All those things point towards a rigorous method that scientists can have faith in. So why does evolution break down?

Because it's not the same sort of science. The science that put men on the moon, the science that built technology, is based upon – you're right – rigorous scientific method. Observation, five senses, accumulation of knowledge. But when you're talking about evolution and Creation, when you're talking about origins, you're talking about history: you weren't there. You are talking about issues in relation to the past, that have no bearing on our technology. I challenge evolutionists, give me one aspect of Darwinian evolution that has anything to do with technology. There is absolutely nothing. The only example they try to give is bacteria becoming resistant to antibiotics and so on, so they are evolving. But that's nothing to do with Darwinian evolution. Certainly, it's to do with mutations and can be due to a loss of information, such as a loss or a corruption of a gene to produce an enzyme, in the *H[elicobacter] pylori* bacterium, which is why then it can't break down the antibiotics, to form poisons to kill it. I think in the penicillin instance, it's an overproduction of an enzyme and so it goes on. But that's where we have to understand, [these are] processes that we can observe right now, and we can understand, that enables us to develop drugs, or whatever, to deal with bacterial resistance. But because mutations are associated with the Darwinian concept, and so on, they say that it's mutating so therefore evolving. And we say not in the Darwinian sense. There are changes, but those bacteria are always bacteria and you can have corruption of genes, you can have loss of information, you can have duplication of information, you can have exchange of information, but it's not Darwinian. For Darwinian evolution you have to have brand new information that never previously existed that's added into the genome, and even Dawkins can't give one example of that.[6]

Jumping 'sideways', do you use microevolution and macroevolution?

We don't tend to use those terms simply because, you know how words change meaning, and certain words mean certain things to the general public. As soon as anyone hears the word evolution, most people would only think of Darwinian evolution. We always like to define the term. So if somebody says 'evolution', we say 'OK, what do you mean?' because the word evolution means change. But the word evolution has become synonymous with molecules to man, so what do you mean? One of the problems as we see it is in the education system. [For example] in the London Natural History Museum (or it used to be anyway in the Darwinian exhibit, but I think they're modifying this exhibit but I haven't seen the new one), but basically they use the word evolution for changes in dogs, or changes in horses or whatever, then they use the word evolution for molecules to man so students are confused as they think evolution is true because changes are happening, which is what you would refer to as microevolution. So to me it's sort of a way in which people are almost brainwashed because they don't understand the difference. It's the same with science, they use the word science when they are talking about the origin of life, which you can't test, you're not there. Then they use the word science for building the space shuttle that put man on the moon. But one involves history, origins that you can't test or observe directly in front of you. The other involves an accumulation of knowledge on the basis of testing an observation which fuels our technology, so we want to make sure we distinguish between those two things.

You mentioned that you have a lot of parallels in Darwinian science and your own understandings. In fact, you embrace a lot of the smaller-scale processes, the temporal processes, which contrasts with certain other understandings. Maybe we could tease out a bit more of that. For example, we have mentioned Kent Hovind's Creation Science, and there's also Carl Baugh's dinosaur footprints.[7] Do you divorce yourself from them perhaps?

Yeah, because, as far as Carl Baugh is concerned, he's a really nice guy and believes the same things we believe but we don't believe there's been [enough] rigorous, scientific investigation done on those particular prints and so on, for us to make any independent comments. So therefore, we would not use it or we keep arm's length from it. We'd have to be careful ... one of the things we've done at Answers in Genesis is we have five PhD scientists on staff, and we have a number that we work with because we want to make sure we maintain the highest of scientific integrity, and not make statements based on a lack of information. But, evolutionists change their models and theories all the time, as we know, and there's a difference between taking a stand on the Bible, and yet we're prepared to change our models built on that. For example, Noah's flood, we're going to say there was a global flood. We'll never change that. But, we will change the model of how the flood occurred – there are scientists doing more and more investigation into things like the accumulation of ice, or the accumulation of sediments, or the erosion of canyons, or deposition strata, how long it takes, fossilisation, all that sort of thing. So, as in evolutionist circles, you might have poor researchers, and you have good researchers. Same is true in Christian circles, let's be honest: there are poor researchers, and those who are good researchers.

That's honest and refreshing to hear. Intelligent Design claims [that] a lot of evolutionary scientists [are] changing their ideas; seeing that there's evidence for methods of design, because they can't see it through natural processes. Is that at the core of your understanding, or is it parallel?

It goes hand in hand with it. See, the Intelligent Design movement is not a Christian movement, as we know. And the Intelligent Design movement actually leaves all of evolutionary history intact. The only thing the Intelligent Design movement really deals with is the issue of naturalism. So they are opposed to naturalism, and we would agree with that. They go on about saying there has to be an intelligence behind life, and we would agree with that. But as Biblical Creationists, we're saying you can't stop there. We want to tell them who the intelligence is, that's the God of the Bible. You see, there are a number of problems if you just say there is intelligence 'out there'.

They are forced to hide their gut feelings of where that intelligence is coming from, so they can try to engage with the scientific community.

Yeah, I understand that, but it doesn't work. Some of them aren't even Christians anyway and are not really on about the Bible. A lot of them believe in millions of years, and even probably evolutionary ideas such as Michael Behe's; I think the best description of him is as a theistic evolutionist, personally. [William] Dembski is certainly an Old Earth Creationist. But the point that we make is that the Intelligent Design movement ... a lot of the general public in America have tried to use it to get Creation into schools, because of the separation of church and state issue, the way it's interpreted by the Supreme Court, and then people say, 'Well, you're really on about the Bible'. And they say, 'No! No! No! We're really on about Intelligent Design'. Well, let's be honest, they are really on about the Bible. For instance, when I was interviewed for the Creation Museum, they said, 'What's your ultimate motive in building a Creationist museum?' Ultimate motive? Well, actually to tell people that the Bible's history is true, and the gospel based on that history is true, and see people saved, and run towards Jesus Christ. Actually, some of the secular press said that it's refreshing that you would admit that. But, we do. And to us, again, if you look around the world you see death, suffering, diseases. If you're just going to say that there is an intelligence who's responsible for those. It's looking through the history of the Bible you understand that death, and suffering and diseases are a consequence of our sin. It's a consequent of God no longer holding everything together perfectly. There's a judgement because of sin. So, we're responsible, not God. So, it's not that intelligence that is responsible for death and suffering, but you only get that from the Revelation in the Bible; if you didn't have that, you wouldn't understand the world correctly. You wouldn't understand God correctly.

One of the real crunches that I think most people are going to have a problem with is the Young Earth aspect. Just the amount of evidence coming from different areas, different disciplines that would point towards a very much older earth than you claim. Again, we're in a great place: James Hutton did a lot of his work and observed the stratification of rocks here…

Ah, but he wasn't a geologist!

No, but Agassiz's Rock is another place where you can go and see these kind of older formations and natural processes. And, you know, you're in Devonian Scotland, which is claiming to be millions and millions and millions of years old.

Something I was going to say earlier, by the way, is for me Darwinian evolution is not really this issue so much as the millions of years. The age of the earth is the crucial issue.

I would agree with that.

Because Darwinian evolution couldn't have taken off without millions of years.

Darwin needs time.

Darwin needs time. And, as I said to someone recently, when I was debating them on radio, in what way does the age of the earth affect your daily living, or your technology or anything like that? Well, it doesn't. And, I said you are very emotional about the age of the earth issue, because you need time for Darwinian evolution, you need time. And then, there's the academic peer pressure. The majority of scientists believe in something, so others, I think, succumb to that. The reason I am a Young Earth Creationist, and zealous with that, is because I'm zealous with the word of God. I would, number one, start with theology. You can't have death and animals eating each other, and diseases like cancer, millions of years before Adam was sent. As soon as you've got millions of years, you do.

That is quite different from the Omphalos hypothesis[8]; I just want to clarify that's not your belief. That seems to be the idea that it's a Young Earth, but God has painted it as an Old Earth to fool man.

I have a problem with Christians who … in fact I had someone who came to me last night – after almost every session someone comes and says to me, 'Don't you think the way to explain it is that God gave earth the appearance of age?', and I said, 'No'. Because that philosophically is a problem. Because then you are admitting it looks old. Then you are admitting that the dating methods work. And I say that the dating methods don't work, they're fallible, the earth doesn't look old. He would have created a mature universe, and a perfect universe, but not with the appearance of age.

Different ways to estimate ages of the universe and of earth: recently, just a couple of months ago, a tight-knit trio of quasars was discovered,[9] millions of years old. So, light

*from stars, isotope decay, carbon dating … all those things you have rote answers for.
Particularly the light from stars. Surely, are these not proven?*

Well, all the things that we say on the lights from stars, in fact, in Creationist research, [this topic] is still very young. We have probably the only PhD astrophysicist-Creationist on staff right now, Dr Jason Lisle, who is actually looking at some of those sorts of issues. But one of the things that he would say, is that Creationists do not have a definitive answer to that, but we would point to those who believe in the Big Bang and say, you have a similar problem.

*That sounds very similar to when Dawkins is speaking – in that he accepts that we don't
have all the answers, but his way is going to find out eventually.*

In a lot of ways, I guess, Dawkins and us are very similar. He's zealous with his faith, and we're zealous with our faith. He seems to spend a lot of time fighting someone that he believes doesn't exist, which is interesting! But, one of the things we say about the light issue, is that from a Big Bang perspective, they have to explain how they can get a light through the whole universe to get that microwave background radiation, and really they only have enough time to get light half way. So, they have a light time travel problem, and we have a light time travel problem. What does it mean? What I would say is, how much do we really know about light? About space? There are so many things we don't know. I guess that's part of the issue, that we believe that evolutionists are not being honest about it. And that's not just in relation to the past – there are all sorts of things that we don't know, and there are all sorts of assumptions that are involved. Whether it's in isotope dating, or whatever it is. So that all those methods are fallible, and there are all sorts of possible things that could happen that you don't know about that can change your conclusions. And then you do get a lot of contradictory data, such as carbon-14 found in diamonds, two billion years old, which they shouldn't be. You've got lots of contradictory data like that.

*Let's jump from rocks to fossils. There's a bit of a giggle among palaeontologists at
the moment, because they found out that the Cincinnatian/Ordovician rock that the
Creation Museum is built on is stacked full of shell fossil deposits, and all of these are
from 490 to 440 million years old.*

According to them.

So, the fallacy is in their dating?

No. When it comes to looking at the fossil record, we have the same fossil record. We have the same order of fossils as they do. We would even agree with them that the older fossils are at the bottom, the younger ones are at the top. But they're probably older by a few months, not older by millions of years.

Because of the depositional rates during the biblical flood?

During the flood and then post-flood catastrophe as well. You know, we would point to all sorts of things. When our geologists are at the Grand Canyon, they look at the layers there and the way they've been uplifted, and bent, and they're not broken and that goes against the idea of millions of years. So, you know, evolutionists also have to look at those things that Creationists bring up, that contradict the idea of a long age, but they tend to ignore those things.

Invertebrate fossils have been found near the summit of Everest.

Yeah, right up at the top.

That's because the flood literally covered all land? How did they get up there?

That's because, even evolutionists talk about mountain uplift, all over the earth. And Creationists would say the same. In fact, if you level out all the mountains and the ocean basins, there is enough water to cover to a depth of two miles.

How did the Himalayas get pushed up in 4000 years?

Well, it was probably a lot less than that.

It goes up by about 5 cm a year.

Right now, but who's to say it wasn't quicker in the past? If there were catastrophic processes. You see, it really comes down to the whole issue of plate tectonics, and so on. In fact, we would say that we would agree that there was probably one continent before the flood. We agree there has been incredible separation of continents. But, we would say it happened pretty catastrophically and in association with the flood. If you look up the work on the website of Dr John Baumgardner, he's got a super computer model dealing with the whole issue of continental split up, continental separation. And you know, basically, one of the things we would say too is like if I was to push this bottle of water, it's quick at first then it slows down. We're right now in the process where it's going slow, but it's been faster in the past.

Another rate thing that people argue against is the post-flood, rapid diversification in animals, two-by-two in the Ark, and the humans are all from one family. Both then experienced a massive explosion of diversity. But because you argue from a genetic, scientific basis, then you must say, 'within kind'; and, that all the potential for variation was there already, 'within kind'. Any clarification on biblical kind?

We would say that in many instances, maybe the majority, but not all, it depends on how things have been classified, but probably more at the family level, would constitute biblical kind. Even evolutionists agree that dogs share a common gene pool, because there has been so much work done on dogs. So the same is true of elephants, probably true of cats – even evolutionists talk about that possibility.

What constrains species 'within kind'?

Part of the information, that's it. I'd say that's the constraining factor. In other words, as I've explained to people from a layman perspective, there's a library of books that builds a dog, and although you can have corruption of that library, there are no new pages or new books that are added.

Evolutionists will argue that because natural selection is all about genes being retained if they are given advantage to reaching reproductive age, and passing it on to the next generation, then every time there is an adaptation, it's effectively new to that organism. So it is, in a sense, new information, about their environment.

It's not new information. You can only have a new combination of information. New combinations can result in some characteristic that can look different.

Isn't there a definition problem there? The information from the evolutionist's point of view is about adapting to a dynamic environment.

Well, what we're talking about is the information in the genome, that we would say, matter can never produce new information, by itself. And of course you can have animals adapting to their environment, there are even mechanisms within certain bacteria. But the interesting thing is that these mechanisms can go back-and-forth, which means they are obviously designed mechanisms, where they can start digesting nylon, which previously they couldn't. There's some information there, that given certain environmental conditions would turn that on and cause that to happen. But it can also turn off, then turn on again, and you see the similar thing, over and over again. So it's obviously, to do with the instructions that are already there.

Geneticists would say the switching works through reading frames and HOX genes, where they get very excited about showing that it is possible to move from, say, an arthropod segmented body to an insect segmented body, and that would be a jump of kind.

Yeah, that would be a jump of kind. I would have to have a molecular geneticist talking about that because I'm not a geneticist.

I'm also out of my depth, but there is a Nature *paper from a few years ago that showed this is possible.*[10]

I might be wrong, I probably shouldn't even comment, but I'm wondering if they're dealing with more homologous genes, and things. Again that's not an example of Darwinian evolution. You can look at similar body patterns then propose some jump but that doesn't mean it happened. I might be wrong, but I think that's what they're doing.

Talking about kind, there are some Creationists who are reluctant to talk about man and apes. I think Ted Haggard had a bit of a go at Dawkins because he was likening his brethren to animals.[11] We are quite similar in our features, our morphology and obviously in our genetics, 96% or so with great apes. But, we're not a kind, no?

> There are a couple of articles on the [Answers in Genesis] website by Dr David DeWitt which specifically deal with those issues. But if there was only 4% difference, even if any of that was true, and again there are some issues with that whole research, but even if it was only 4% then you're talking about three billion base pairs. Four per cent is millions and millions and millions of differences. The other thing is too, what they are starting to realise now, is that they feed into a whole other level of information that they are just not aware of in the cells; they believe there is information outside of the nucleus [epigenetics]. There are all sorts of things that they are talking about there. What about all the junk genes? They no longer believe they are junk genes. The latest research is that they are not junk genes at all. So what about if you started counting all those? So I think we've got to be really careful with that sort of stuff. I'd say, you'd expect similarity across the animal kingdom, and even with man, who I separate out, because we're breaking down sugars, or we're doing things with oxygen – there are all sorts of metabolic processes that we all have similar, so you would expect similar instruction; but there are also major differences as well.

There's a new Darwin bicentenary exhibit at the Answers in Genesis Creation Museum. It even presents the view that natural selection can coexist with Young Earth Creationism. However, in case there is any doubt, a large sign states: 'Natural Selection is not Evolution'.

14

THE DISSENT OF MAN

Gin he be spar'd to be a beast

Like Owen, we may be proud and supercilious about being different from animals, thanks to our brain development, and we may even attribute our differences to the craftiness of some Supreme Being. And yet we have more in common with animals than not. Ironically, it was the study of homology, in Owen's own field of comparative anatomy, then physiology and now genetics which have all revealed our many similarities with the other animals. Is it possible for us to marvel at that without undermining our superiority, and while still retaining our higher intelligence? K'ung Fu-tzu (Confucius) observed: 'Mankind differs from the animals only by a little, and most people throw that away'.

Bizarrely, the opposite desire, to reveal differences within the Hominidae, has actually led to confusion over classification of humans and non-humans. A case in point is the very first scientific study of a man-like ape, known as 'Tyson's Pygmy',[1] a misidentified immature Angolan chimpanzee dissected by Edward Tyson (1650–1708), who became the founder of comparative anatomy when he compared its anatomy with those of man and apes. He concluded: 'Our Pygmie is no man, nor yet the Common Ape; but a sort of animal between both'. Tyson foreshortened this distance 'between' by depicting the animal upright and holding a walking stick. His motivation was to position his pygmy on one of the rungs of the ladder that links all animals in a God-given and static hierarchy, the 'great chain of being'. Man, of course, is at the top of this chain, with the pygmy somewhat lower down, but still above the apes. However, the real revelation here is that Tyson did recognise enough similarity to suggest a link between man and ape, albeit inadvertently. But it was over 200 years later before Darwin did so.

Our relatedness to the great apes, and animals at large, has been a constant source of dissension from religious quarters, especially Christianity. This objection over descent has even produced an irony where strength of feeling about human superiority, and our God-given rights, actually led to a human wronged. Ota Benga was of the Batwa people in the Congo. Having survived the massacre of his village, including his wife and two children, by the Belgian army, he was sold to an American businessman in 1904 and shipped back to be displayed alongside an orang-utan in the monkey house at the Bronx Zoo, thereby causing quite an uproar. This was not least because, like Tyson's Pygmy, the attempt to place the Batwa pygmy on the evolutionary scale between humans and apes was obvious. An African-American Baptist clergyman protested: 'Our race, we think, is depressed enough, without exhibiting one of us with the apes [...] We think we are worthy of being considered human beings, with souls [...] The Darwinian theory is absolutely opposed to Christianity, and a public demonstration in its favor should not be permitted' – so precipitating Benga's eventual release from the zoo, but only after the keepers had him as a roving exhibit to which the public's prods and verbal abuse would be met with monkey business, and ultimately violence. Benga's fate was to be Westernised, suffering the ignominy for a decade, before eventually committing suicide.[2] The point here is our relationship with the apes, not that religion led to this man's suffering. That was unfortunate, and however misguided, the Baptists' intentions were righteous.

Darwin would have sympathised with Benga's plight. He was famously anti-slavery, having been disgusted at the treatment of Brazilian slaves. His egalitarianism no doubt also stemmed from his freethinking background, with the whole family tended towards Unitarianism. Additionally, the ultimate impact on Darwin of having mixed with similarly open-minded naturalists at Edinburgh cultivated in him an interrogative psyche that was not evident beforehand, and one that was to remain with him forever after. Darwin benefited from the parallel teaching of both theory and practice that would resonate throughout his future life's work. This, coupled with his lack of prejudice, formed within him a powerful capacity for looking at the world with a rare clarity of vision.

Scientific theory and practice are brought together through the way science looks at the world, by applying a hypothetico-deductive method of interrogation. Hypotheses are developed through theorising and philosophical reasoning, often by making predictions based on logic. Hume called this *a priori* knowledge the Relations of Ideas. Experiments are then designed to test hypotheses, providing observations for analysis which produces evidence for, or against, the truthful accuracy of the hypothesis. If the accuracy is acceptable, ideally the hypothesis can be accepted as theory and becomes *a posteriori* knowledge, and what Hume called a Matter of Fact. Darwin clearly understood Hume's empiricism. His son Francis wrote about how his father approached scientific work, a couple of years after his passing:

> He often said that no one could be a good observer unless they were an active theoriser [...] it was as tho' he were charged with theorising power ready to flow into any channel [...] but fortunately his richness of imagination was equalled by his power of judging & condemning the thoughts that occurred to him [...] and so it happened that he was willing to test what would seem to most people not at all worth testing. These rather wild trials he called 'fool's experiments' & enjoyed extremely [...] his wish to test the most improbable ideas. This wish was very strong in him and I can remember the way he said 'I shan't be easy till I've tried it' – as if an outside force were compelling him.

'[H]e was *willing* to test.' Without this state of mind, Darwin may well have never questioned the traditional understanding, reinforced in his time by Natural Theology, and without his explanation of evolutionary processes he would never have reached the uncomfortable but inevitable contradiction of the widely accepted biblical account of human origins. Darwin's conclusions were especially uncomfortable on a personal level. They were a direct threat to his wife's piousness, which produced in him a terrible conflict between his professional intent and the protection of her feelings.

If Oscar Wilde wasn't far off the mark when he wrote, 'Life imitates art far more than art imitates life',[3] then there is, within a small exedra in the Ashworth Laboratories at the University of Edinburgh, a piece of art, a famous statuette, that beautifully captures Darwin's well-documented struggle. It is the *Affe mit Schädel* (Ape with Skull) by the German sculptor Hugo Rheinhold. A duplicate takes pride of place at the Aberdeen Medico-Chirurgical Society.

The *Affe mit Schädel* depicts a Common Chimpanzee (*Pan troglodytes*) sitting atop a higgledy-piggledy pile of books and manuscripts. The subject is likely female, based on her smaller brow ridge, and in the absence of any contradictory, genital evidence. She has in her right forepaw a human skull, echoing the scene in Shakespeare's *Hamlet* where the Prince of Denmark mourns Yorick ('Alas, poor Yorick! I knew him...'). She cradles her chin with her other forepaw in a contemplative posture. Her left foot holds the shin of her right leg as if to steady it or support the callipers held by her right foot. A spine of one of the closed books reads 'DARWIN'. The book open at her feet and facing the viewer has a single inscription on the right-hand page, 'ERITIS SICUT DEUS'. This is a quote from Genesis 3:5, when the serpent is enticing Eve to eat of the tree of knowledge, promising, 'And ye shall be as God [knowing good and evil]'). However, the second half of this quote is missing, ripped from the lower half of the page.

What inspired Rheinhold in making his sculpture is not known for sure. It has obvious parallels with Auguste Rodin's *The Thinker*, but it is perhaps surprising to discover that they are not likely related; while Rodin had developed his statue as early as 1880, it was not cast into bronze and displayed until after the *Affe mit Schädel* had debuted at the Great Berlin Art Exhibition, in 1893.

A large part of the statuette's popularity is the myriad of possible interpretations. A message explicitly made elsewhere by Rheinhold cautions against imprudent use of technology. In the case of the *Affe mit Schädel*, the excised biblical quote suggests that good and evil cannot be known, or distinguished. With the ape's study, the library of books and the calliper instruments, the suggestion is that the statuette is warning against the application of rationalism in the absence of morality. The *Affe mit Schädel* also deals with mortality. Specifically, when a human is depicted holding a skull it is usually about the inevitability of death. But it is something quite different for our ape, who is engaged in assessment and measurement. The countenance is not one of sorrow or melancholy, but studious indifference, or even whimsy.

One might hazard that the statuette in part symbolises Darwin's own studious career: his engagement with the question of human origins, his empiricism, the exacting inspection of his own ideas informed through his assessment of other literature, and ultimately his own realisation of the conflict between evolution and Creationism that forced so much personal soul-searching. A scholarly, non-human hominid acts as a powerful reinforcement of our primitive ancestry, and Darwin's role in elucidating our evolutionary past. In this sense, the *Affe mit Schädel* is not so much Darwinian as *it is Darwin*.

Rheinhold's *Affe mit Schädel*.

Photograph © Steven Hay 2010

Notes and artefacts from Darwin's time spent in Edinburgh

Earlier in this book are letters Charles Darwin wrote to his family and friends while in Edinburgh. He writes about his time spent in Scotland, or as a consequence of his interaction with Scots. His correspondence in almost its entirety can be viewed at *The Darwin Correspondence Project* (http://www.darwinproject.ac.uk/).

Few other documents have survived from Darwin's student days at Edinburgh, but those that have are preserved in the Cambridge University archives and are being made available via *The Complete Work of Charles Darwin Online* (http://darwin-online.org.uk/) project, in which his private papers and those unpublished by Darwin are denoted 'DAR'.

DAR 129 contains a diary of zoological and botanical observations made in 1826, particularly of his time spent walking with his brother Erasmus.

DAR 271 is a recently unearthed gem: Darwin's 'Edinburgh Reading List', thought to date from about January 1826, but could also have included some items added retrospectively.

DAR 118 is known as his 'Edinburgh notebook' and shows a marked development in his recording of observations since his diary entries from the preceding year, likely through the influence of Grant. Importantly this archive also records his observations on *Flustra* and *Pontobdella muricata*, soon after contributing to his first published scientific paper which he read to the Plinian Society in March 1827.

DAR 5 contains some of his lecture notes of 1825–6, on medicine and chemistry, an account of a zoological walk to Portobello, along with other miscellanea, such as his Plinian Society session ticket, and some notes he made on Lamarck's classification of invertebrates.

DAR 130 is Darwin's 'Glen Roy notebook' containing his geological observations from Edinburgh, especially Salisbury Craigs, and Glen Roy in 1838.

The following section presents examples of these notes and diary entries written by Charles Darwin while in Edinburgh between February 1826 and April 1827, and upon his return in 1838.

Diary extracts for 1826 with entries about birds, beasts and flowers seen on walks recorded in Darwin's 1826 A.W. McLean's 'The Edinburgh Ladies & Gentlemens Pocket Souvenir Diary', dedicated to 'The Nobility and Gentry of Edinburgh, This little Work is respectfully inscribed, by their most obedient, and most faithful Servant, the Publisher (* Early preparation shall be taken with the Souvenir for 1827, to render it worthy of their notice)'.

The Edinburgh
Ladies & Gentlemens
POCKET SOUVENIR
FOR
1826.

Edina! Scotia's darling seat! All hail thy Palaces and Towers.

Lizars sc.

EDINBURGH:
PUBLISHED BY A.W. McLEAN
Register Street.

Erasmus caught a Cuttle fish it had a bill like a Parrots & near it a bag of of black coloured fluid: is the little fish which emits when pursued renders turbid in water an English species? yes

Is it the Sepia Loligo? Yes now the Loligo sagitalla

15

Caught an orange coloured globular (Zoophite?) was fixed to a rock & when kept in a bason would turn itself inside out & when touched retracted itself in again; much in the same way as a glove is turned inside out; put it in spirits:

Of note, be records on 24 Feb 1826, 'Bought a Ptarmigan', possibly for his taxidermy sessions with John Edmonstone.

February.　　1826.

13 MONDAY.

14 TUESDAY.

15 WEDNESDAY.

Darwin's diary shows that he and Erasmus would go searching the rock pools between Leith and Portobello.

[27 Feb–1 March 1826]

27

In the pools of water left by the sea there were a great many roundish–conical Actenia? of a bright red colour firmly fixed to the rocks; when kept on a plate they turned themselves inside out & could entirely change the shape of their bodies

Are they the Actinia crassicornis or mesembryanthemum.

[2–4 March 1826]

The shore was literally covered with little fish, when touched they emitted a dark coloured fluid & I think eve on seeing any body coming their process of swimming is extremely curious. they first inflate themselves with water & then fixing their tentacula on the sand. at this same slightly bending their bodies send forth the water to distance of three or four feet with considerable noise. & it seems by the reaction that they first put themselves in motion. they thus proceed

[6–8 March 1826]

with considerable rapidity. their tail being the only part exposed.— they swim tail foremost & N.B. This is very uncommon suspension think it some event among the fishes

7

Saw three Snow Buntings shot. they were flying in small flocks about tech shore. one of them a great deal whiter & more beautiful than the rest is this the Cock or Hen?

Found a common star fish with only three arms. the other two having been torn off. two new ones half an inch long were just beginning to grow. —

His success in encountering shore-life continued a few days later, when on 10 March 1826 he records, 'Caught a very large Sea mouse – Size 5 1/2 inches. are no uncommon on the shore between Leith & Portobello'. also noting, 'A great many Sea mice on the shore. when thrown into the sea rolled themselves up like hedgehogs'.

There are no entries in the diary for November and up until 23rd December 1826 when Darwin records, 'Saw Grey Wagtail & Water Ouzel under Braid Hills', and then . . .

[25–27 December 1826]

25

A remarkably foggy day. So much so that the trees condensed the vapour & caused it to fall like large drops of rain

Saw a hooded Crow feeding with some rooks. by the sea shore. near Leith.

It's of interest that he spent Christmas in Edinburgh. Was it not celebrated at the family home in Shrewsbury, or did he simply prefer to stay away?

Edinburgh Reading List

Books that I have read thro since my return to Edinburgh[1].
Franklins Journal to the North. Pole[2]. 2 Vol. 8 Vo.[3]
Cochrane Travels in Columbia[4]. 2 Vol. 8 Vo
Abernethy Physi. Lectures[5] 1 Vol.
Scoresby account of Polar Regions[6]. 2 Vols. 8 Vo.
Darwins Zoonomia[7] 2 Vols 4to[8].
Paris Pharmacologia[9] 2 Vols 8 Vo
Pamplets by D[rs]. Grant & Brewster on Nature History[10]. 7 in number
Blairs lectures on Belles Lettres[11]. 3 Vol. 8Vo
Abernethy Hunterian orat & Lect. Pamph[12]. 8 Vo.
H K Whites Letters & Poems[13] 12 mo[14]
Pennants Arctic Zoology[15] 2 Vols 4to
Several papers in the Werner. Trans.[16]
Several numbers in the New Edinb: Philos Journal[17].
Bostocks Physiology[18] 2 Vol 8 Vo.
Cuviers theory of the earth[19]. 1 Vol. 8 Vo
Almack[20] & Granby[21] 6 Vols 12. mo
Henry Chemistry[22] 2 Vols 8 Vo
Sewards memoirs of Darwin[23] 1 Vol 8 Vo.
Several essays in Rambler[24]
Brambletye House[25] 3 Vos 12 mo
Clarkes travels[26] 5 Vols. 4to.

---

1 In October 1826 CD returned to Edinburgh for the second year of his medical studies (see letter from E. A. Darwin, [29 September 1826]).
2 Franklin 1823.
3 8 Vo: octavo.
4 Cochrane 1825.
5 Abernethy 1822. There is a lightly annotated copy in the Darwin Library--CUL, bound with Abernethy 1819a, Abernethy 1819b and Abernethy 1823. The volume has "Erasmus Darwin" on the flyleaf.
6 Scoresby 1820. There is an annotated copy in the Darwin Library--Down
7 E. Darwin 1794--6. CD's annotated copy is in the Rare Books Department--CUL.
8 4 to: quarto.
9 Paris 1825. There is a copy in the Darwin Library--Down.
10 CD probably refers to the \Edinburgh journal of science\ (1824--32), which was edited by David Brewster (\ODNB\); Robert Edmond Grant contributed several papers on invertebrates (\ODNB\; Desmond and Parker 2006).
11 Blair 1790.
12 Abernethy 1819a; CD perhaps refers to Abernethy 1819b and Abernethy 1823 (see n. 3).
13 White 1826.
14 12 mo: duodecimo.
15 Pennant 1784--5.
16 \Memoirs of the Wernerian Natural History Society\ Vols 1-8i (1808-39).
17 \Edinburgh new philosophical journal\ vols. 1--19 (1826--64), a continuation of the \Edinburgh philosophical journal\ vols. 1--14 (1819--26).
18 Bostock 1824--7. Volume 1 is in the Darwin Library--Down.
19 Jameson trans. 1827. There is an annotated copy in the Darwin Library--CUL.
20 Hudson 1826.
21 Lister 1826.
22 Henry 1823. Volume 2 is in the Darwin Library--CUL.
23 Seward 1804. There is a copy in the Darwin Library--Down.
24 \Rambler\. 1--208 (1750--2). A periodical by Samuel Johnson.
25 Smith 1826.
26 Clarke 1810--23.

Bibliography to the Edinburgh Reading List

Abernethy, John. 1819a. *The Hunterian oration, for the year 1819: delivered before the Royal College of Surgeons, in London.* London: Longman, Hurst, Rees, Orme, and Brown.

Abernethy, John. 1819b. *Part of the introductory lecture for the year 1815, exhibiting some of Mr. Hunter's opinions respecting diseases, delivered before the Royal College of Surgeons, in London.* New edition. London: Longman, Hurst, Rees, Orme and Brown.

Abernethy, John. 1822. *Physiological lectures, exhibiting a general view of Mr. Hunter's physiology, and of his researches in comparative anatomy: delivered before the Royal College of Surgeons, in the year 1817.* 2d edition. London: Longman, Hurst, Rees, Orme, and Brown.

Abernethy, John. 1823. *Introductory lectures, exhibiting some of Mr. Hunter's opinions respecting life and diseases: delivered before the Royal College of Surgeons, London, in 1814 and 1815.* New ed. London: Longman, Hurst, Rees, Orme, and Brown.

Blair, Hugh. 1790. *Lectures on rhetoric and belles letters.* 4th edition. 3 vols. London: A. Strahan and T. Cadell; Edinburgh: W. Creech.

Bostock, John. 1824–7. *An elementary system of physiology.* 3 vols. London: Baldwin, Cradock and Joy.

Clarke, Edward Daniel. 1810–23. *Travels in various countries of Europe, Asia and Africa.* 3 pts in 6 vols. London.

Cochrane, Charles Stuart. 1825. *Journal of a residence and travels in Colombia, during the years 1823 and 1824.* 2 vols. London: H. Colburn.

Darwin, Erasmus. 1794–6. *Zoonomia; or, the laws of organic life.* 2 vols. London: J. Johnson.

Desmond, Adrian, and Sarah Parker. 2006. The bibliography of Robert Edmond Grant (1793–1874). *Archives of Natural History.* 33: 202–13.

Franklin, John. 1823. *Narrative of a journey to the shores of the Polar Sea, in the years 1819, 20, 21, and 22.* London: John Murray.

Henry, William. 1823. *The elements of experimental chemistry.* 9th ed. 2 vols. London.

Hudson, Marianne Spencer. 1826. *Almack's: a novel.* 3 vols. London: Saunders & Otley.

Jameson, Robert, trans. 1827. *Essay on the theory of the earth with geological illustrations.* By Georges Cuvier. 5th ed. Edinburgh: William Blackwood. London: T. Cadell.

Lister, Thomas Henry. 1826. *Granby.* 3 vols. London.

ODNB: Oxford dictionary of national biography: from the earliest times to the year 2000. (Revised edition.) Edited by H. C. G. Matthew and Brian Harrison. 60 vols. and index. Oxford: Oxford University Press. 2004.

Paris, John Ayrton. 1825. *Pharmacologia.* 6th ed. London: W. Phillips.

Pennant, Thomas. 1784–5. *Arctic zoology.* 2 vols. London: Henry Hughs.

Scoresby, William. 1820. *An account of the Arctic regions, with a history and description of the northern whale-fishery.* 2 vols. Edinburgh: Archibald Constable and Co., Edinburgh, and Hurst, Robinson and Co., London.

Seward, Anna. 1804. *Memoirs of the life of Dr. Darwin.* London: J. Johnson.

Smith, Horace. 1826. *Brambletye House: or, cavaliers and roundheads.* 3d ed. 3 vols. London: Colburn.

White, Henry Kirke. 1826. *The life and remains of Henry Kirke White of Nottingham: late of St. John's College, Cambridge.* London: J. F. Dove.

The Edinburgh Reading List: transcription, footnotes and bibliography reproduced by kind permission of the Darwin correspondence Project at Cambridge University.

Edinburgh notebook

March 16th 1827.—

(1) Procured from the black rocks at Leith a large Cyclopterus Lumpus (common lump fish). Length from snout to tail 23 ½ inches, girth 19 ½. It had evidently come to the rocks to spawn & was there left stranded by the tide; its ovaria contained a great mass of spawn of a rose colour. Dissected it with Dr Grant. — It appeared very free from disease & had no intestinal worms; its back however was covered with small crustaceous animals. — Eyes small. — Hence probably does not inhabit deep seas? Stomach large. Liver without gall-bladder. Kidneys situated some way from the Vertebrae: an unusual fact in cartilaginous Fishes. — Air bladder was not seen. Brain very small; the optic nerves being nearly as large as the spinal cord, neither the brain or spinal matter

2)

nearly filling its cavity. — The valves in the heart were very distinct; the peduncle strong. The body was not covered with skirscales, but slimy & remarkably thick. The sucker on its breast was of a white colour. I believe it is generally a reddish yellow? The plebs differ whether it is edible. —

(2) Procured a small green Aolis1 & a Tritonia.

(3) Examined the ova of the Purpura Lapillus & found them out of their capsules & of this shape

18th

(4) Found these growing out of an Alcyonium.—?

(5) Some ova from the Newhaven rocks said to be that of the Doris, was in every respect similar to that of the Univalves & in rapid motion, & continued so for 7 days.

19th

(6) Observed ova in the Flustra Foliacea & Truncata, the former of which were in motion. —

I may mention that I have also observed ova of the Flustra Foliacea & Truncata in motion. That such ova had organs of motion does not appear to have been hitherto observed either by Lamarck Cuvier Lamouroux or any other author: — This fact although at first it may appear of little importance, yet by adducing one more to the already numerous examples will tend to generalise the law that the ova of all Zoophites enjoy spontaneous motion.

—

This & the following communication was read both before the Wernerian & Plinian Societies.

Session ticket to the Plinian Society

Grant facilitated Darwin's engagement with evolutionary ideas: 'He one day, when we were walking together burst forth in high admiration of Lamarck and his views on evolution.' Here Darwin makes notes on Lamarck's classification system for invertebrates.

M. Lamarck arranges "Les Animals sans Vertebra" into 7
orders. viz:

  1.ˢᵗ Les Mollusques
  2.ᵈ Les. Crustacées
  3.ᵈ Les. Arachnides
  4.ᵗʰ Les. Insectes
  5.ᵗʰ Les. Vers
  6.ᵗʰ Les. Radiaires
  7.ᵗʰ Les. Polypes

The first of which he thus subdivided
  Mollusca with head.
    1. Naked
  a. Those which swim at liberty. as Sepia
  b. Those which creep on ye belly as Limax
    2. Covered with shell
  a. One celled. not spiral as Patella
  b. One celled spiral. umbilicate as Buccinum
  c. One celled. imperforate. as Turbo
  d. manycells melong ye animal as Nautilus
    Mollusca headless
    1. Naked. as Ascidia
    2. Covered with shell
  a. Two valves equals with or without acc. one. as Mytilus
  b. two valves unequal inclines in tube. as Teredo.

A second page ends this list with:

c. Two unequal valves with or without hinge, as Ostrea
d. …two valves as Balanus

The latter being a genus of barnacles and a subject Darwin would return to at length.

Finally for this section, here is an immediately recognisable Edinburgh landmark: Darwin's sketch of Salisbury Craigs made during his return in 1838.

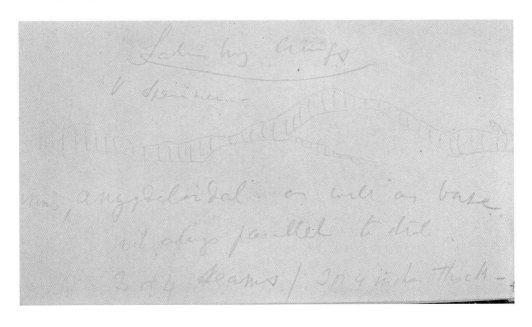

Salisbury Craigs
V. Specimens —
Veins, amygdaloidal — as well as base not always parallel to strata
3 or 4 seams / 3 or 4 inches thick —

15

PSYCHOLINGUISTICS

The mair they talk, I'm kent the better

Our language is often identified as the main difference between us and other animals.[1] Verbal communication is not at all unique to humans, but in all its varied forms, it does seem more complex than grunting, barking, hooting, honking, bleating, cawing, mewing and neighing, and all those other vibrations emanating from the animal kingdom. Well, that complexity may very well be why the human brain has become the biggest that there is, relative to body size, and looks the way it does,[2] and possibly the reason for its rapid bursts in growth and development. Supporting evidence for this comes from other social functions, like face recognition, being controlled by a part of the brain that also exhibits some language activity.[3] Thus, language and social behaviour very likely co-evolved but, and perhaps surprisingly, some experts consider that monogamous love in pair-bonding has been most important in shaping our brains.[4]

However, psycholinguistics famously falls roughly into two schools of thought about language evolution. Skinner's 1957 version of verbal behaviour (now morphed into relational frame theory) proposed learning through a Pavlovian conditioned response as brain capacity, intelligence and society gradually co-evolved. Language was thus considered a naturally selected Darwinian adaptation. This is in contrast with Chomsky's innate universal grammar from 1965, a set of linguistic structures hard-wired into our brains. This may have evolved as a spandrel,[5] a by-product of another adaptation, perhaps one involved with repeated problem-solving-in-society. This recursive ability of the mind could have then been borrowed from its original context, leading to a neuronal reorganisation, and the rise of syntax. Of importance, this process would not have needed natural selection. This also means that it could have been a comparatively speedy process[6]; in *The Selfish Gene*, Dawkins included language as part of human culture, and therefore an example of memetics: 'Language seems to "evolve" by non-genetic means, and at a rate which is orders of magnitude faster than genetic evolution'.

The voice inside your head, the internal dialogue part of our language, may be a prerequisite of thought itself.[7] Between editions of *The Descent of Man*,[8] Darwin became aware of ideas that 'the use of language implies the power of forming general concepts' and that '[t]here is no thought without words, as little as there are words without thought'. On language evolution, he seemed to have a foot in both psycholinguistic camps, suggesting evolutionary development of language as well as instinctual elements: 'I cannot doubt that language owes its origin to the imitation and modification of various natural sounds, the voices of other animals, and man's own instinctive cries, aided by signs and gestures [...] may not some unusually wise ape-like animal have imitated the growl of a beast of prey, and thus told his fellow-monkeys the nature of the expected danger? This would have been a first step in the formation of a language [...] The formation of different languages and of distinct species, and the proofs that both have been developed through a gradual process, are curiously parallel'. Given that linguistic universals would need to have developed prior to language diversification, Noam Chomsky is understandably non-committal about interpreting language evolution in Darwinian terms:

> I don't separate Darwin from language evolution. He did. He didn't try to address the problem seriously, which is no criticism. [...] The formation of different languages is something that happened *after* the evolution of the shared human language capacity, hence has essentially nothing to do with evolution of language (apart from what it teaches us about the genetic capacity, shared among humans). As for distinct species, Darwin – notoriously – had very little to say about it, apart from some suggestions that have been absorbed into the theory of speciation, which is almost entirely post-Darwin, in fact rather modern. [...]
>
> Darwin has no notion of gradualism in language, apart from a few scattered sentences that could be accepted in almost any approach that is vaguely within the framework of biology [...] The influence of Darwin is that everyone seriously interested in biology (and in my view for the last 50-odd years, language should be studied as part of biology) takes for granted that natural selection is a major factor in evolution. And in some specific areas, that insight has led to very significant achievements. In other areas it has not. Evolution of cognitive capacities is, for the most part, one of these areas. The reasons are pretty clear: the fossil evidence is very slight, and the archaeological evidence thin. So it is necessarily mostly speculation. [I] think there is a possibility of serious work on evolution of language.
>
> Contemporary humans are very similar genetically. They apparently separated about 50 000 years ago, and even apart from interaction, that is far too short a time for any significant evolutionary process to have taken place. Cultural

differences are certainly relevant to language research; that's why linguists try to study as many languages as possible. But they are not relevant to research into evolution of language, for the reasons mentioned, except in the indirect sense that language variety tells us a lot about the shared genetic capacity that evolved before the trek from Africa about 50 000 years ago.

While consensus appears to be in favour of some form of pre-adaptation for language capacity, there is still a debate over the subsequent changes that must have taken place, especially those involved with grammatical structure. Ideas revolve around the gradual emergence of a social commentary, especially between detached groups, and modification of languages via cultural transmission from generation to generation. What is clear is the enormous potential for complexity arising from interaction of individual learning, cultural transmission and biological evolution, all adaptive and operating across different timescales, and constrained by learning bottlenecks during childhood.[9] With this in mind, Simon Kirby would push the biological language envelope even further, but within cautionary limits:

Over the last few decades there has been growing acceptance of the idea that language can be viewed from an essentially biological perspective – that our species has a unique biological endowment which enables the acquisition and processing of language. This view of language as a species-specific faculty, along with the rise in popularity of evolutionary approaches to psychology in general, has led many researchers such as Steven Pinker to adaptationist explanations for the origins and structure of language. The idea is that if language is a complex biological trait and if it appears to be adapted to some function (such as the communication of complex meanings), then natural selection provides the most appropriate explanatory mechanism.

However, this view is not universally accepted. Indeed, a relatively recent development in theoretical linguistics has been the idea that the uniquely human aspect of language may not be as complex as previously thought (see, for example, work by Noam Chomsky and colleagues). If this is the case, then what is special to humans may not after all require a complex adaptive explanation. Nevertheless, this view actually highlights those aspects of our language faculty that we may share with other species, thus opening the door to evidence from comparative biology. Whereas linguists have traditionally dismissed the relevance of animal communication to the study of human language, I strongly suspect that future linguistic inquiry will be increasingly informed by what we have learnt from other species. Of this, I am sure Darwin would have approved.

Another recent development in evolutionary thinking that has influenced the study of language has been the popularity of looking at cultural evolution from a Darwinist perspective. One of the most striking features of human language is that it is to a large part culturally transmitted. When a sentence is uttered, it not only conveys semantic information but also information about the particular language of the speaker. Children use this information in combination with their biologically given language faculty to 'reverse engineer' the language of the speech community they are born into. When they themselves speak, this process is repeated with the next generation of language learners. There are clear parallels here between the transmission (and therefore evolution) of cultural and genetic information over time. Although we must be careful not to stretch the analogy too far, there have been a number of attempts to show how selectionist mechanisms may apply to the cultural transmission of language. Treating language itself as an evolutionary system may help explain not only how it emerged, but also ongoing language change that is visible today.

Finally, and most unexpectedly, Darwinian ideas have provided linguists with a new tool for modelling the origins and evolution of language. For much of the last century, computer scientists dreamt of the possibility of building life-like processes *in silico*. Taking inspiration from the simple underlying principles of neo-Darwinism, they saw life as an essentially computational process and therefore amenable to simulation (or even replication) on digital computers. As computing resources increase in speed and decrease in cost, it has become quite feasible to simulate populations of communicating and evolving individuals either on the computer or even in robotic experiments. This has opened up a whole new framework within which evolutionary linguists can test their theories, particularly when looking at ways in which learning, culture and evolution interact. Just as developments in artificial intelligence gave us new ways to look at cognition, so too do evolutionary computation techniques help us understand complex evolving systems such as language.

So there are several routes along which Darwinian ideas have travelled to converge on the central questions of linguistics, such as why language is the way it is, and how our species came to possess it. When I reflect on the influence of Darwin on my field, however, I realise it is less to do with the specific ways in which we can take an evolutionary stance on language. In the end these may not be particularly 'Darwinian'. Rather, I am struck by Darwin's remarkable insight that simple dynamic processes can nevertheless give rise to emergent complexity. In demonstrating the success of this insight, his influence on the study of *all* complex systems cannot be underestimated. Ultimately, he inspires us to take the step from description to explanation.

16

INTELLIGENT DESIGN

O thou great unknown Power!

The human intellect is desperate to investigate the context of its very existence, but there are huge gaps in our understanding. That's okay; most people would probably think it honest and honourable to admit to not knowing everything. But what about claiming that you know a certain amount and then invoking a supernatural explanation for the bits you don't? To then use this in argument is like turning your own ignorance into a weapon against your adversaries. The Intelligent Design movement, the most recent case attempting to suggest wholesale doubt within scientific ranks, unfortunately has the appearance of doing just that. The scientists largely refuse to be drawn to debate, stating that Intelligent Design is, ironically, just a resurrection of Paley's Natural Theology, and thus a modern twist to Old Earth Creationism. This section is in two parts: in this chapter we deal with some biological and theological aspects of Intelligent Design, and in the next chapter we investigate some political ramifications, specifically the consequences for our educational systems.

Agnosticism has become a bit of a dirty word for being non-committal in the arenas of strongly held views and bullish opinion. Well, there was nothing irresolute about Huxley who introduced the term in 1869 to describe his firm rejection of Natural Theology while still being able to accept biblical moral teachings. Like his friend and colleague Darwin, Huxley also strove for truth and understanding through Hume's scientific method; he considered Hume's *Natural History of Religion*, one of the first robust naturalistic analyses of faith as a human behaviour, to have 'anticipated the results of modern investigation'. Thus, it was not a cowardly escape by Huxley, to admit to there being gaps in his understanding: without (Greek: *a-*) knowledge (Greek: *gnosis*). It was a confident and courageous salute to the complexity in nature, and complexity in our intelligence, as a part of nature. From Huxley's later essay *Agnosticism* in 1889:

> When I reached intellectual maturity and began to ask myself whether I was an atheist, a theist, or a pantheist; a materialist or an idealist; Christian or a freethinker; I found that the more I learned and reflected, the less ready was the answer; until, at last, I came to the conclusion that I had neither art nor part with any of these denominations, except the last. The one thing in which most of these good people were agreed was the one thing in which I differed from them. They were quite sure they had attained a certain 'gnosis', – had, more or less successfully, solved the problem of existence; while I was quite sure I had not, and had a pretty strong conviction that the problem was insoluble. And, with Hume and Kant on my side, I could not think myself presumptuous in holding fast by that opinion [...] So I took thought, and invented what I conceived to be the appropriate title of 'agnostic'. It came into my head as suggestively antithetic to the 'gnostic' of Church history, who professed to know so much about the very things of which I was ignorant; and I took the earliest opportunity of parading it at our Society [...] Agnosticism, in fact, is not a creed, but a method, the essence of which lies in the rigorous application of a single principle. That principle is of great antiquity; it is as old as Socrates; as old as the writer who said, 'Try all things, hold fast by that which is good'; it is the foundation of the Reformation, which simply illustrated the axiom that every man should be able to give a reason for the faith that is in him; it is the great principle of Descartes; it is the fundamental axiom of modern science. Positively the principle may be expressed: In matters of the intellect, follow your reason as far as it will take you, without regard to any other consideration. And negatively: In matters of the intellect do not pretend that conclusions are certain which are not demonstrated or demonstrable. That I take to be the agnostic faith.

Hume's empiricism demands that an idea must be demonstrable before becoming a known Matter of Fact, and Huxley's agnosticism was an admission to gaps in that knowledge. He was making a move away from Creationism, whereas Intelligent Design is seen as a political 'God of the gaps', and a return to Creationism. The tragedy is that Intelligent Design advocates are honestly trying hard to integrate science and religion, ulterior more political motives aside. The travesty is that it is presented purely as a science, whereas in fact it doesn't fully match Huxley's agnosticism as an exacting and discerning methodology for truth.

However, one of the main advocates, William A. Dembski, defends Intelligent Design as having a starting point quite different to that of Creationism:

> Creationism is always a doctrine about where did everything come from ultimately. Intelligent Design really falls under the engineering sciences. What we found is in the last 30 years with advances in molecular biology and biochemistry, and the information sciences, is that there are features of biological systems that cannot be understood and explained apart from

through intelligence, or purpose. What we have in the cell, for instance, we've got signal transduction circuitry, complex motors, miniature machines. We need nanotechnology and engineering sciences to understand these explicit systems, and explain how they came about in the first place ... what is the ultimate intelligence behind things? It could be an alien intelligence, it could be a natural intelligence built into the world.[1]

Although this argument is well versed in the language of modern science, it does echo the incredulity voiced by Paley in his *Natural Theology* in which he famously includes his watchmaker's analogy:

In crossing a heath, suppose I pitched my foot against a stone, and were asked how the stone came to be there; I might possibly answer, that, for anything I knew to the contrary, it had lain there forever: nor would it perhaps be very easy to show the absurdity of this answer. But suppose I had found a watch upon the ground, and it should be inquired how the watch happened to be in that place; I should hardly think of the answer I had before given [...] There must have existed, at some time, and at some place or other, an artificer or artificers, who formed [the watch] for the purpose which we find it actually to answer; who comprehended its construction, and designed its use. [...] Every indication of contrivance, every manifestation of design, which existed in the watch, exists in the works of nature; with the difference, on the side of nature, of being greater or more, and that in a degree which exceeds all computation.

Of course, Paley named God as his creator of complexity, something from which Intelligent Design supporters tend to dissociate and leave open to question. Rather, the argument for complexity is made solely on the basis of our current explanations being unsatisfactory; for William A. Dembski, Darwinian natural selection is incapable for the purpose:

Before Darwin, the power of choice was confined to designing intelligences – to conscious agents that could reflect deliberately on the possible consequences of their choices. Darwin's claim to fame was to argue that natural forces, lacking any purposiveness or prevision of future possibilities, likewise have the power to choose. Accordingly, Darwin invented an oxymoron: natural selection. In ascribing the power to choose to natural selection, Darwin perpetrated the greatest intellectual swindle in the history of ideas. Nature has no power to choose. All natural selection does is narrow the variability of incidental organismal change by weeding out the less fit. Moreover, it acts on the spur of the moment, based solely on what the environment at the present time deems fit, and thus without any prevision of future possibilities. And yet this blind process, when coupled with another blind process, namely, incidental organismal change (which neo-Darwinians understand as genetic mutations), is supposed to produce designs that exceed the capacities of any designers in our experience. No wonder Daniel Dennett, in *Darwin's Dangerous Idea*, credits Darwin with 'the single best idea anyone has ever had'. Getting design without a designer is a good trick indeed. But with advances in technology as well as in the information and life sciences (especially molecular biology), it's a trick that can no longer be maintained. It's time to lay aside the smokescreens and the handwaving [*sic*], the just-so stories and the stonewalling, the bluster and bluffs, and explain scientifically what people have known right along, namely, why the appearance of design in biology is not merely an appearance but in fact the result of an actual intelligence. This is the fundamental claim of Intelligent Design.

The immediately obvious question at this point, and one asked many times over, is what is the source of that designing intelligence? Darwinists already have a solution, but what is the alternative? Daniel C. Dennett makes a suggestion:

In the decade since *Darwin's Dangerous Idea* appeared, I have spent thousands of hours discussing the implications of Darwinism with people in the arts, humanities and social sciences, and I never cease to be amazed at the complacency with which so many of them just assume that they can dismiss out of hand any Darwinian account of human learning or creativity. What do they think is going on in their brains, and how do they imagine that brains work their 'magic'? They are sure that something wonderful is going on, no doubt, something much too complicated for them to understand, but they ignore the analysis that cuts through all this complexity and points so unmistakably in Darwin's direction: A learning or innovating brain is – by definition, really – a brain whose current design enables it to revise its own design in appropriate response to novel information. Self-redesigning design processes! Is such a thing possible? Either brains have been designed by non-miraculous means to revise their own designs by non-miraculous means, or minds are miracles. Darwinian natural selection is the only model of such processes that is known to work, and it works at many levels. People who would rightly scoff at anyone who imagined there were flying carpets or perpetual motion machines need to confront their own breezy assurance that minds are neither miraculous nor Darwinian in their origin, development, and operation. Can they describe an alternative?

Perhaps not an alternative, but claims of evidence for design are in no short supply. The human eye is often cited by anti-Darwinists as an example of miraculous natural complexity. A frequent part of the Darwinist defence suggests the eye's imperfection is a historical record of its evolution: the vertebrate eye may be considered to be built backwards and upside down, compared with the alternative, more optimal, design of the cephalopod 'camera eye'. However, George E. Marshall puts this argument down to anatomical ignorance:

I have recently changed disciplines from cell engineering to anatomy. Assuming that Darwinism can be equated with evolution from molecules to man, I find the concept quite irritating from an anatomical point of view. In my experience, when anatomists don't understand the function of a structure, they explain it off in evolutionary terms. The appendix is a classic example. What irritates me is that it closes the door to exploring possibilities to its function. In this context, evolution is a blinker that shuts down scientific enquiry. It is more profitable to assume that everything has a function and that nothing is wasted. Let us be man enough to admit that we don't know and not just sweep it under the carpet of evolutionary throwbacks. Unfortunately this has more serious implications than simply academic and I cite one case to illustrate.

The curve of the lumbar spine toward the front (lordosis) is unique to man among the primates. It was thought by evolutionists to be the problematic result of man having recently adopted an upright position. Some researchers blamed back pain on this, saying that the spine had not yet evolved satisfactorily. However, as Professor Richard Porter assumed that it must be designed for its function, he realised that the arch of the spine, like the arch of a bridge, adds strength. A person with a lumbar lordosis can lift proportionally more weight than a gorilla with its kyphotic (curves toward back) spine. This realisation revolutionised treatment, as treatments which restore the supposed evolutionary disadvantaged lordosis work exceedingly well. Evolutionary thinking seems to breed the assumption that biological structures are not optimally designed. Professor Richard Dawkins, for example, suggested that the eye was 'wired the wrong way round', an assumption admittedly that I made when looking at the structure of the eye as a 13-year-old school boy. It took a couple of years of studying eye anatomy before the penny dropped. Let me explain. The light-detecting structures within photoreceptor cells are located in a stack of discs. These discs are being continually replaced by the formation of new ones at one end of the stack and old ones being 'swallowed' by a single layer of retinal pigment epithelial cells (RPE) at the other end of the stack. RPE cells are highly active, and for this they need a very large blood supply – the choroid. Unlike the retina, which is virtually transparent, the choroid is virtually opaque, because of the vast numbers of red blood cells within it. For the retina to be wired the way that Professor Dawkins suggested would require the choroid to come between the photoreceptor cells and the light, for RPE cells must be kept in intimate contact with both the choroid and photoreceptors to perform their job. Anyone who has had the misfortune of a haemorrhage in front of their retina will testify as to how well red blood cells block out the light.

Undoubtedly, at least to the non-specialist, Marshall appears to have formed a scientific conclusion to inform his faith. Another high-profile case receiving quite a lot of media coverage in 2004 involved philosopher Antony Flew at the University of Reading, who switched from atheism to deism partly for what he perceives as evidence for Intelligent Design in DNA. These two examples stand out because of the authority of their scientific arguments for God. Professional opinion is clearly given a lot of credence for publicity purposes, as indicated by the use of petitions for Intelligent Design campaigns. An announcement made in 2001 by the Discovery Institute, the largest body funding the development of Intelligent Design, claimed that scientists were bravely coming out of the anti-Darwinian closet to stage a challenge by questioning convention, 'A Scientific Dissent From Darwinism'. A few hundred signed a petition stating that '[w]e are skeptical of claims for the ability of random mutation and natural selection to account for the complexity of life. Careful examination of the evidence for Darwinian theory should be encouraged'. The number of signatories and their credentials are unimportant; the petition was dramatically dwarfed by the tens of thousands of scientists who signed several petitions in response, including the catchily titled 'A Scientific Support For Darwinism And For Public Schools Not To Teach "Intelligent Design" As Science'.

The Discovery Institute, and, less so, John Calvert's Intelligent Design Network, have associations with many of the main proponents of Intelligent Design and their publications, from the early incarnations of the biology school textbook *Of Pandas and People*, up to the most recent, Ben Stein's high-profile film *Expelled*. Irrespective of the format, the argument from Intelligent Design largely boils down to the issue over the capacity of Darwinian evolution, and specifically natural selection, to explain complexity. Michael J. Behe, renowned for his design analysis of the bacterial flagellum, believes that the explanatory power of Darwinism is often only assumed:

> To gauge Darwin's profound impact on molecular biology compare the following two sentences, the first from a recent journal article: (1) 'Every cell has evolved mechanisms that identify and eliminate misfolded and unassembled proteins'. (2) 'Every cell has mechanisms that identify and eliminate misfolded and unassembled proteins'. Now ask, how do the two sentences differ? What information does the missing verb add to the first sentence? Here's another sentence from the same paper: 'To accomplish a timely recognition of nascent membrane proteins by the translocon-bound RNC [ribosome nascent chain complexes] during the integration process, the system has therefore evolved a mechanism that utilizes ribosome-induced folding of selected nascent chain sequences within the ribosomal tunnel'. Now re-read that sentence, leaving out the words 'therefore evolved'. What tangible knowledge was lost? Isolated examples these are not. If one is attuned to it, everywhere in the literature one encounters the utterly gratuitous use of the word 'evolved'.
>
> Darwin didn't just propose a theory, he bequeathed a view of nature. That view requires nature to behave in certain ways, which most of us now take for granted. Another giant of science who for a while had a similar impact was James Clerk Maxwell, whose elegant electromagnetic theory supported the wave nature of light. Of course waves required a medium in which to propagate. Writing in the ninth edition of the *Encyclopaedia Britannica* Maxwell confidently asserted that 'there can be no doubt that the interplanetary and interstellar spaces are not empty, but are occupied by a material substance or body, which is certainly the largest, and probably the most uniform body of which we have any knowledge' – the ether. Yet the amount of actual, direct evidence 19th-century physics mustered to support the existence of the ether was zero – the same as the amount of direct evidence 21st-century biology has that Darwin's theory explains 'mechanisms that identify and eliminate misfolded and unassembled proteins'. One measure of the power of an idea is its ability to make even brilliant people overlook its unsupported premises.

It's not surprising that such an emotive subject often becomes clouded by vitriol. But what remains unclear are the boundaries between Darwinism and the Intelligent Design understanding and acceptance of evolution *per se*. Antony Latham is not an explicit advocate of Intelligent Design, but has come to realise that he has followed an independent and personal path of understanding, through various fields of education, that has converged on conclusions that are very similar to Intelligent Design (hence recognition of his work in the USA). He has also campaigned for debate of Intelligent Design in Scotland, and has strongly held views on objectors to such moves, Richard Dawkins in particular:

The four major concepts of Intelligent Design are Irreducible Complexity, Specified Complexity, Fine-tuned Universe and Designer. A Fine-tuned Universe implies that there is a formulation or a structure that's predetermined to enable this to happen. Your book, 'The Naked Emperor: Darwinism Exposed', goes through an extraordinary range of contents, and you start by mentioning Martin Rees.

> Well, it's really to set the scene of Design. He writes very elegantly about the fine tuning of the universe in a book called *Cosmic Habitat*, which I read and have quoted from quite freely, because I know that he is a very authoritative and good scientist. He comes up with a lot of the facts to do with the very exquisite fine tuning of many constants and factors from the Big Bang onwards. And, basically, we're talking about such fine precision, to the 10th decimal point. This is for many things – for instance, the expansion energy of the universe and the gravitational force pulling it back at the first moments. That's just one example. Then, if you move it minutely, either way, then you don't get galaxies, you don't get the universe as we know it, you don't get life. And so basically, to me, and to many, not just Christians, but to many cosmologists and others, this is an indication of a Designer. Certainly that is one way of looking at it, because it's so perfectly set up. The way that I think [Rees] copes with it, because I don't think he's a Christian, and many others do, is they postulate multiple universes. So, we've just happened to strike lucky out of many possible, perhaps billions of universes, and that allows them to cope with the improbability that we are in the one that we're in.

> [...] if you're going to talk about Intelligent Design, you need to account for convergence: what does it mean in an evolutionary context? The ones that I've picked up are the camera eye, which I've found in about seven different groups of organisms, and are supposedly to have evolved separately, but they're very similar, like the octopus eye, the human eye, they're not the same but they're similar. Another good one is the electrical fish, who use an electromagnetic field to sense their environment and these seem to have evolved completely separately, but almost precisely the same, in both the Old and New Worlds [...] ultimately apart from the microevolutionary level, which I've no problem with. I'm not a Darwinist on the macroevolutionary level.

Compared with some of the major exponents of Intelligent Design who don't want to package it as such, you're very strongly Christian-based.

> Well, I think quite a number of them are actually Christians. I mean, Behe's a Catholic and Dembski's a Christian [...] I've been uninfluenced in that sense by the American Intelligent Design people. I wrote a book that I felt should have a bit of a personal journey in it. In the Introduction, I say why

I'm even interested in this whole thing [...] I explain in the Introduction that I had a belief in God as a child, you might say simplistic. Basically, brought up in a church-going family. But, as soon as I came across evolution in secondary school, in biology class, then I realised that ultimately, the basis of that process was random: the mutations are the driving force of variety. And almost unconsciously my faith disappeared, well at least I became an agnostic, not an atheist. I used to enjoy to some extent arguing with Christians, and people who came up to me, and I'd say, 'Well actually, come on, it's random. There's no evidence for any Divine hand in life, or why we're here at all'. I know a lot of people don't ever see it like that, but to me that was clear. It's maybe the way I think. It was just an inevitable conclusion that if Darwinism is correct, then it does almost throw Christianity out of the window: the implication is that it doesn't require a Divine hand.

The 'God of the gaps' idea talks about how more and more gaps in our knowledge are filled with new discoveries in science, and how God is eased out.

Yes ... [but] we know so much more now, particularly about the molecular side of cell biology, that far from the gaps getting smaller I would suggest that they're getting bigger.

So was this actually through educating yourself, or education, that you found out more information that then put God back into the gaps?

No, [...] I became a Christian quite outside of all this stuff and I don't actually go into how I became a Christian, but ultimately it was God revealing himself to me through reading the Bible, and so on, when I was in Kenya actually, when I was working out there. And it was only after that that I realised I had this conviction that we are created in some fashion. I certainly wasn't a Young Earth Creationist. I just hadn't a clue basically. So I wanted to find out really what is the evidence, and look at all the evidence as objectively as possible. I had admittedly not a very advanced biological training, up to sort of A-level standard and then I did medicine and so on. So, when I had a chance about four years ago, I went part-time for a bit in my doctor work, and I enrolled in an Open University course on evolution. I went through that course, I didn't actually complete, didn't actually do all the assignments, I wasn't a very good student. I did some of them. I became so engrossed in reading the massive textbook, and then I just became absolutely fascinated apart from anything else.

Dembski mentions your book on his website as the first studious literature from the UK on Intelligent Design.

It is implicit in my arguments [...] I think that it could be put down as Intelligent Design. I'm quite happy for it to be called that [...] I'm not a Young Earth Creationist, and I find that the argument is clouded very much by the Young Earth Creationists, and get quite annoyed when I see the documentaries and stuff where people like Dawkins will immediately latch onto the extreme people and rubbish them, which I mean, they're open to that, however genuine they believe their case [...]

The biggest chapter in the book is a critique of *The Blind Watchmaker*, and I go into a bit about the eye, which I find incredibly simplistic. It's very shallow I believe, his argument. I think all his arguments are quite shallow, actually! I have nothing against Dawkins as a person, and admire him as a writer, and I admire his conviction even, and his determination. I think where I disagree with him on the eye process ... he says, imagine a series of X's, one slightly different from the next, slightly more improving, this photosensitive area becomes cupped, and then it forms the membrane and then you get a lens. He says it's really like almost each little stage is hardly different from the previous stage; he's a pure Darwinist, he can't cope with sudden leaps, it must be very, very, very gradual otherwise there must be something else going on which is not random. And where I disagree with him is that it's not a matter of small quantitative changes, it's qualitative changes that you see in the eye. You've got at some point to get a pupil, for instance. Now, that is an extraordinary thing, and it's not just a matter of tinkering with what is already there, it is something that has to be, in a Darwinian sense, it has to be useful, otherwise it won't be selected. And you've got to try to imagine what the first pupil did, and how it worked.

The same goes for the lens: I believe that you can't just get any old blob of stuff hanging in the eye. It's got to have a noticeably beneficial quality otherwise it won't be selected, and it's got to fit in with all the genetic stuff that's already there, which is working. And he talks about one organism, a *Nautilus*, one of these cephalopods, which actually doesn't have a lens. And he actually admits that he almost loses sleep at night wondering why it hasn't evolved a lens, because, you know, a lens would help it so much. And I challenge him, saying how can you imagine any mutation giving it a lens? I ask him, and anybody else, how would a mutation, overnight, which it has to be for the first organism to have a proto-lens that works, how is it going to do it?

[...] he says that you can't just get a haemoglobin molecule with 146 amino acids, appropriately ordered: it has to happen in small stages. He uses a line of Shakespeare, 'methinks it is like a weasel', and he gets his daughter to type on to a computer randomly to see in how many generations you can get this sentence. Now, where the whole argument falls completely and utterly flat on its face, is that in the computer he has the full sentence, and the computer recognises when his daughter hits the right key. So when she does, [hits] an 'm' at the beginning, the computer thinks, 'okay' and slots the 'm' in, and keeps it there. And it's a totally un-Darwinian idea. You cannot have a bacteria that knows it wants to be a human [...] He's put into the computer advanced knowledge of the sentence, and that is not Darwinian [...] If he hadn't put the final sentence as the template to go for, then he would still be doing it, it would not happen. It's only because he falsified it by putting in the final sentence that he is able to get there in only a few generations. It is very straightforward and totally false, and people accept it. Unbelievably, people think that he has written a good book! It's utterly wrong and false. I get upset about it, partly because he is so strong on his atheistic stance.

[...] It's almost unbelievable that an academic biologist would do such nonsense. He has this computer which has this branching algorithm, and he looks at it, and he himself is the person to decide, so he is already putting intelligence into the thing, by just doing that. That's the first error in his reasoning. The next thing, is that it's just so banal and naïve, and simplistic. You've got a three-dimensional branch and he says, 'Wow! Looks like a beetle'. Honestly! I know it's popular science but he's trying to reach and influence people, and I find that just outrageous. I am amazed that he is in his position. He uses that computer simulation, these morphs, and it's nonsense. It's unscientific, pie in the sky. I'd prefer to go back to Darwin, he was a real scientist.

A facet of Intelligent Design is Specified Complexity: Dembski says that a single letter of the alphabet is specified without being complex, a random sentence is complex without being specified, a Shakespearean sonnet is both complex and specified, and therefore the details of living things can be similarly characterised, especially patterns of molecular sequences in functioning biological molecules (e.g. DNA), and that these could not happen through chance.

Complexity on its own could happen by random. If it is a specified complexity it has a specific function, and form which is fitting for its role. Therefore, it is not just complex, but it's specified. I'm not a mathematician, but Dembski used maths to show that can't happen. One of the revelations for me is the difference between micro- and macroevolution. It would be very wrong to say that I am totally un-Darwinian. Basically I don't have a problem with microevolution, the reshuffling and recombination of existing alleles or genes. It's a fact in breeds of dogs by artificial selection, Darwin's finches, the cichlids in Lake Victoria, and even the melanic moth, where there was clearly selection going on during the industrial era where a certain allele conferred a darker colour. But that doesn't explain macroevolution, because you're talking about existing genetic material which definitely does have an enormous potential for variety within it. You only have to look at dogs for that, even though they are still the same species, it's quite extraordinary. And they seem to share the same genome as a wolf virtually.

There is within a group or species or genus a wide potential for variety, because there are so many alleles for each gene, and therefore microevolution can very easily work on that. The faster cheetah will survive, and therefore its offspring will have the propensity to be faster. But, it never explains brand new qualitative organisms. I'm okay about speciation, if you want to call Darwin's finches different species. Up to that point, Darwin was right. He obviously had the fossil record, but he struggled over that and the gaps in it, saying that it was an imperfect record. But, his main observations were to do with domestic breeds, pigeons, dogs and the varieties of plants. And beautiful, wonderful observations that he made. It was inevitable, without knowing anything about the real genetics going on, that he would assume that this was enough to provide the variation and the fodder for selection to work on. But we now know that isn't enough, that these variations that he was looking at are existing genes, in alleles that are in the genetic pool of that species, and this is microevolution [...] it does look as though there are two things, micro- and macro-.

121

[...] The Open University book merely says, that as far as macroevolution [is concerned] – brand new organs, wings, feathers, whatever it is – that these are due to mutations, random events in the genetic code which then somehow confer advantages in a few cases. That's where I start to say Darwinism falls down. I'm very fond of Darwin; I have a chapter all about his life, and I feel so sad about Darwin because I feel he was a genuine, quite humble, sort of chap. I think he went through a lot of knocks, his daughter died, and the whole thing shattered his Victorian faith in the benign God. But I try to be as objective as possible in the book: can mutations give you a trilobite eye, for instance? Or a lens, or an iris?

Can you tell us how the macroevolution works, involving the Designer?

I have no idea, and I try to say that at the end. It is an utter mystery. But I don't think it's unscientific. This is where I will differ from many critics of the whole thing. I don't think it's unscientific to be talking about the possibility of miracles. Some of the greatest scientists believed in miracles, Newton and Einstein. To the likes of Dawkins, it's just rubbish and it explains nothing, and you have to explain God, and how can a complex being have come about? And I try to briefly tackle that as well. Miracle is miracle, and I don't think you can ever understand miracle. Now, the pure scientist in the 21st century baulks at that. You are outside science, and you shouldn't even be inside science, to be talking like that. I would argue, no, I think that's right. I think that's a very recent phenomenon to say that everything has to be by natural laws.

There's so much evidence that things are not just by natural laws, be it the fine tuning of the universe, to the complexity that we see in life. The open-minded scientist will be able to sit comfortably with his quite rational understanding of how things normally occur, alongside the possibility of Divine input as well [...] I think it's very hands-on [...] I don't think we're ever going to find out how He did it, and that's where I think scientists, I believe, need to have some humility about it. We have done marvellously well in working things out, but there comes a point where we may not understand how macroevolution occurred, and what exactly happened.

[...] I don't take Genesis literally. I see it as a beautiful allegory. I think key to Genesis is that humans are undoubtedly special and part of Creation and in line from animals, with humans at the end. Although Genesis is not a scientific text, it tells the truths as you need to know, that we were created in some way, and that the universe was created from nothing, and that God was there beforehand. Humans are undoubtedly special and somehow made in His image.

[...] I show an agnosticism in my book about whether there is literally a lineage from a common ancestor with chimpanzees, for instance, to humans, through *Australopithecines* and *Homo erectus*, etc. And this would go along with homology – vertebrate limb which is similar in all vertebrates, be it the whale flipper or human arm, a wonderful, powerful argument for evolution. I don't have a problem if there's some form of descent, which would have involved precursors, which you can see in the fossil record definitely, a progression from simple to complex, and humans appearing on the very end. But, I don't think it happened in such a perfect, gradual, random variation of form, and the evidence is against that. And so the saltationist view, which I hold, implies some extra information being put in, to create human beings, in an incredibly short time, evolutionarily speaking.

And at which stage has there been this enormous Divine input? At the Neanderthal stage? At an earlier stage?

The fact is, is that modern *Homo sapiens*, as far as we know, are 100-to-200 000 years old. And, I don't know. Obviously we lived with Neanderthals in Europe for thousands of years but seem to have been separate. And my main showing in the book is the improbability, or impossibility I would say, of the tripling in size of the human brain in three million years, with relatively few generations. After all, we're not *Drosophila*! Or bacteria. We take 20 years for a generation, and if you go back five million years that's only 200 000 generations or something, which is not that many considering the enormous changes that have to occur in the brain and everything else. Also, I try to show the difference between any other animal, ape or otherwise, from humans as regards complexity of language. I think the language is possibly the thing you can home in on, quite apart from general intelligence, which separates us from every other beast.

17

SCHOOL OF LIFE

Taught him great Nature's Magna Charta

There have been consolidated moves to introduce Intelligent Design into the science curriculum in the USA. The nationwide fervour in response is partly a function of the American separation of church and state, and their vigilance over this First Amendment. However, simultaneous action could be possible across several states because there is no national school board for setting curricula, so that US states can set their own teaching standards and the school curriculum can be devised right down to the individual district. But, as seen in Pennsylvania, it was the individual members of the Dover Board of Education who determined that particular outcome: the school board voted 6–3 that, from January 2005, teachers would be required to read out this statement to their ninth-grade biology students at Dover High School:

> The Pennsylvania Academic Standards require students to learn about Darwin's theory of evolution and eventually to take a standardized test of which evolution is a part.

> Because Darwin's Theory is a theory, it is still being tested as new evidence is discovered. The Theory is not a fact. Gaps in the Theory exist for which there is no evidence. A theory is defined as a well-tested explanation that unifies a broad range of observations.

> Intelligent design is an explanation of the origin of life that differs from Darwin's view. The reference book, *Of Pandas and People*, is available for students to see if they would like to explore this view in an effort to gain an understanding of what intelligent design actually involves.

> As is true with any theory, students are encouraged to keep an open mind. The school leaves the discussion of the origins of life to individual students and their families. As a standards-driven district, class instruction focuses upon preparing students to achieve proficiency on standards-based assessments.

Entrenched views are said to be reinforced by the professional and social groups to which we belong. Given the national tendencies, the comprehensive resistance by the teachers to this proclamation possibly came as quite a surprise: for example, a survey in 2007 recorded that 47% of US High School biology teachers believed in God-guided evolution, compared with 28% believing in an unguided process. A further 16% were Young Earth Creationists. The equivalent to a US High School education in the UK is provided at secondary level. The 2001 UK census[1] revealed that of all of the 'Secondary education teaching professionals', 69% considered themselves Christian, with about half as many more women than men, and 22% had no religion. Although comparing these figures assumes atheism goes hand-in-hand with an acceptance of evolution, it is striking how similar the percentages are for the USA and UK.

That's secondary school teachers only, but comparing the 75% of all people calling themselves religious, in response to the 2001 UK census, with the 18% who did not, which professions would you place in each upper quartile (top 25%)? To make it really easy, here are the two lists in descending order, topmost professions first:

> List A is: clergy, school secretaries, farmers, school mid-day assistants, school crossing patrol attendants and medical secretaries.

> List B is: psychologists, social science researchers, artists, higher education teaching professionals, arts officers; producers and directors, non-educational researchers, journalists; newspaper and periodical editors, physicists; geologists and meteorologists, software professionals, graphic designers, conservation and environmental protection officers, broadcasting associate professionals, archivists and curators, authors; writers, photographers and audiovisual equipment operators, IT strategy and planning professionals, musicians, countryside and park rangers, scientific researchers, natural environment and conservation managers and architects.

List A was, of course, the upper percentile of professions worked in by religious respondents, either Christian, Buddhist, Hindu, Jewish, Muslim or Sikh, and including 'all other religions'. List B was therefore the people responding 'no religion', which is a little different from purely atheists or agnostics and is definitely not a vote for evolution,[2] but it still illustrates the stereotypical expectation, and no surprises there: you are more likely to be a Darwinist if you are some sort of scientist or technical type (which must also account for the artists), and a Creationist if you are not.[3] In this way it seems that affiliations conspire to maintain divisions, but the big question now is, if our career paths are in part dictated by our education, then what is happening in our schools?

Before the Intelligent Design movement is able to make itself heard about what they present as their scientific arguments, they have to engage with the politics of schooling, and this is where they argue from a different manifesto. The US legal system assumes that religion and science are mutually exclusive, so instead of identifying specific gaps in Darwinism to imply a supernatural intervention, Darwinism is portrayed as 'a theory in crisis', for which their answer is 'teach the controversy'. This is a view that was shared by President George W. Bush in 2005: 'I think that part of education is to expose people to different schools of thought [...] you're asking me whether or not people ought to be exposed to different ideas, the answer is yes' – although, during the 2000 presidential election campaign, he had already shown himself to be somewhat partisan on the subject: 'I don't necessarily believe every single word [of the Bible] is literally true, I think that, for example, on the issue of evolution, the verdict is still out on how God created the earth'.

The call to 'teach the controversy' has received little sympathy where tested. The famous 2005 *Kitzmiller* v. *Dover Area School District* trial ruling concluded that the status of evolutionary science had been misrepresented and that '[Intelligent Design]'s negative attacks on evolution have been refuted by the scientific community'. Indeed, the aspersions cast on Darwinian science probably had the opposite effect to that intended. Largely in response to the education-oriented promotion of Intelligent Design, mainly by the Discovery Institute, the US National Center for Science Education (NCSE) launched 'Project Steve' as 'a tongue-in-cheek parody of a long-standing Creationist tradition of amassing lists of "scientists who doubt evolution" or "scientists who dissent from Darwinism" '. Their petition statement reads:

> Evolution is a vital, well-supported, unifying principle of the biological sciences, and the scientific evidence is overwhelmingly in favor of the idea that all living things share a common ancestry. Although there are legitimate debates about the patterns and processes of evolution, there is no serious scientific doubt that evolution occurred or that natural selection is a major mechanism in its occurrence. It is scientifically inappropriate and pedagogically irresponsible for Creationist pseudoscience, including but not limited to 'intelligent design', to be introduced into the science curricula of our nation's public schools.

Quite surprising because it seems so out of character, one of the most famous US Darwinists, Daniel C. Dennett, has spoken out on the subject of education, advocating his own, all-inclusive, curriculum for US schools:

> [...] on facts about all the religions of the world. About their history, about their creeds, about their texts, their music, their symbolisms, their prohibitions, their requirements. And that this should be presented factually, straightforwardly with no particular spin to all of the children in the country. And, as long as you teach them that, you can teach them anything else you like. That, I think, is maximal tolerance for religious freedom. As long as you inform your children about other religions then you may, and as early as you like, and whatever you like, teach them whatever creed you want them to learn, but also let them know about other religions [...] Because democracy depends on an informed citizenship.[4]

Dennett and the most famous UK Darwinist, Richard Dawkins, class themselves as like-minded colleagues, and so it's perhaps no surprise that they share similar goals towards freedom of choice when it comes to religion – although, at first glance, Dennett's idea for schooling seems far more autonomous than Dawkins' calling religious indoctrination and education of minors 'mental child abuse':

> How can you possibly describe a child of four as a Muslim or a Christian or a Hindu or a Jew? Would you talk about a four-year-old economic monetarist? Would you talk about a four-year-old neo-isolationist or a four-year-old liberal Republican? There are opinions about the cosmos and the world that children, once grown, will presumably be in a position to evaluate for themselves. Religion is the one field in our culture about which it is absolutely accepted, without question – without even noticing how bizarre it is – that parents have a total and absolute say in what their children are going to be, how their children are going to be raised, what opinions their children are going to have about the cosmos, about life, about existence. Do you see what I mean about mental child abuse?[5]

However, even if topical, this concern over religious indoctrination is nothing new. It can also be found in the roots of social reform, and a globally celebrated case for this was famously made in Scotland. The industrial revolution seemed to rise out of the technological advances made simultaneous to the Age of Enlightenment. Industrialists were on the whole tyrannical, and workers' conditions desperate. Fortunately a few in management were visionary, and went about manufacturing a better future.

Robert Owen (1771–1858) married into a mill-owning family, took part ownership of the New Lanark mill and purpose-built village on the River Clyde, about 35 miles southwest of Edinburgh, and proceeded to manage it under a new set of higher principles based on education, child care and elderly care, social inclusion, good parenting, equality, medicine and community responsibility. About 500 of the 2000 employees were children brought in at the age of five or six from the poorhouses and charities of Edinburgh and Glasgow. Owen established a school, the world's first infant nursery, a shop with healthy fresh produce and an 'Institute for the Formation of Character'. He believed that a person's character was formed through external influences for which they were not responsible, and therefore could not be held to blame:

> [N]o infant has the power of deciding what period of time, or in what part of the world he shall come into existence; of whom he shall be born, in what particular religion he shall be trained to believe, or by what circumstance he shall be surrounded from birth to death.[6]

Owen also considered that religion had historically been used as a tool to control people and to keep them oppressed, sentiments more usually associated with Karl Marx (1818–1883) and his statement that 'Die Religion [...] ist das Opium des Volkes'[7]: 'Religion is the sigh of the oppressed creature, the heart of a heartless world, and the soul of soulless conditions. It is the opium of the people. The abolition of religion as the illusory happiness of the people is the demand for their real happiness'. Except that, compared with Marx's mostly academic career, Owen was actually in a practical position to implement his version of Utopian Socialism:

> [...] when religion is stripped of the mysteries with which the priests of all times and countries have invested it, and when such is explained in terms sufficiently simple that the common mind can fully comprehend it, without fear or alarm from a misguided imagination, all its divinity vanishes; its errors become palpable; and it stands before the astonished world in all its naked deformity of vice, hypocrisy, and imbecility. If, indeed, religion, as it has been hitherto taught, has emanated from a divinity, it must have been one possessing the most dire hatred to mankind; one, who well knew how the most effectually to destroy, in the bud, the finest qualities of human nature, and the highest enjoyments of every individual [...]

> In order to govern through the influence of the religion which has been described, it becomes absolutely necessary, first to make the people mental slaves, and to devise institutions, by which their rational faculties shall be perverted or destroyed from infancy.

It is worth noting that this lecture was given in 1830, and that Owen died the year before the *Origin of Species* was published, and so social objections to religion already existed long before biological justification for having them became foremost. Today it is often the opposite, where social objections are appended to support naturalist arguments.

Compared with the USA, and in the absence of any landmark court cases such as in Pennsylvanian Dover, the UK seems more relaxed about religious education in schools. In fact, religious tuition is a compulsory part of the curriculum, set out in guidelines for the Curriculum for Excellence (CfE) pedagogy. Consequently, our local primary school in Edinburgh follows the national programme of Religious, Moral and Philosophy Studies (RMPS). Considering the quarter of people under the UK census who registered themselves areligious, plus all the other non-Christians, it is most likely that many parents would welcome their children being taught a balanced, even secular, view of their world. Ideally their programme would fit Dennett's well-informed society model, and it might make for a more tolerant society in future.

These are certainly among the ultimate aims of the UK Campaign for Secular Education whose understanding of Scottish education in 2008 was as follows:

> Altogether there are 2183 schools administered by 33 local councils acting as education authorities. Most schools are non-denominational (read

128

Protestant), with 391 denominational (read Catholic) that were established to accommodate the Irish immigrants; there's one Jewish school and three Episcopalian. It also looks as though there will be one Islamic school soon because the First Minister [...] is a very religious man (Catholic) and is more than sympathetic to the idea. In general, local clergy visit every school, sometimes regularly [...]

In primaries, the RME [Religious and Moral Education]/RO [Religious Observance] provision is combined making it a more difficult nut to crack because kids only have one teacher for all subjects. So, the amount of time given to promoting religion really depends on that teacher's (or Head's) worldview – and, most importantly, how (s)he interprets the guidelines. So, according to the beliefs of the teacher then, religion can either be taught on a 'some people believe' model, or the Christian model taken for granted with rituals and stories permeating each and every subject.

Understandably, this interpretation seems rather dramatic and hypothetical, but it does ring true from personal experience. The local school certainly considers itself 'non-denominational' which means that their RMPS seeks to use 'a variety of themes to develop an awareness of Christianity and Other World Religions. To encourage positive attitudes towards other people and to the environment. To develop an awareness of some basic moral values'. These are admirable aims, if fully realised and if the presentation of religions is balanced, even accepting some slight, but inevitable, bias towards Christianity. Sadly this seems not to be the case, and the reality is less comprehensive, insidious perhaps, with only nominal nods being made to Diwali and a few other aspects of non-Christianity, and only within the scope of RMPS. In contrast, and for two years running, a preliminary class in grammar for six-year-olds has as an example of verb use, 'the Good Samaritan *crossed* the road', and on another occasion, the homework exercise was to reorder and find as many words as possible within 'God made families'. Additionally, in my youngest daughter's recent annual report, without mention of any other religious text, we are told that she 'understands that the Bible is a special book and is becoming familiar with some of the stories in it'.

Intelligent Design in the UK has yet to catch up with the USA. There are some developments. Most recently, in 2007 the organisation Truth in Science was blocked by the UK Government from disseminating Discovery Institute material in 'information packs', including lesson plans, directly to UK schools. This was in response to an e-Petition, 'to prevent the use of creationist and other pseudo-scientific propaganda in Government-funded schools'. A Government minister at the Department for Education and Skills said that 'Neither intelligent design nor creationism are recognised scientific theories and they are not included in the science curriculum; the Truth in Science information pack is therefore not an appropriate resource to support the science curriculum', in keeping with the official Government line, released on 19 June 2007:

> The Government is clear that creationism and intelligent design are not part of the science National Curriculum programmes of study and should not be taught as science. The science programmes of study set out the legal requirements of the science National Curriculum. They focus on the nature of science as a subject discipline, including what constitutes scientific evidence and how this is established. Students learn about scientific theories as established bodies of scientific knowledge with extensive supporting evidence, and how evidence can form the basis for experimentation to test hypotheses. In this context, the Government would expect teachers to answer pupils' questions about creationism, intelligent design, and other religious beliefs within this scientific framework.

It's not clear how this accords with the UK Government's funding of 7000 state faith schools, including 37 Jewish, seven Muslim and two Sikh. Instead of a consistent approach, in a rather hypocritical move, the funding is to be increased for more schools, particularly Muslim ones, to 'give the Government greater control over Muslim schools at a time when questions are being raised about whether some are adequately preparing children for life in Britain'.

Meanwhile, an online survey of more than a thousand teachers claimed that a third of them disagreed with the Government's position on Creationism and Intelligent Design not being part of the science National Curriculum. Nine out of 10 were sympathetic towards Michael Reiss, the former director of education at the UK Royal Society, who resigned in September 2008 over comments about including Creationism in science lessons, agreeing that they would discuss religious matters, including Intelligent Design, in science lessons if pupils brought them up.

Latest, as of March 2009, is that in 70 secondary schools across the English county of Hampshire, '[t]eachers are being given advice on how questions about evolution and God can be explored with 11- to 14-year-olds', in science and religious education lessons. The president of the National Secular Society responded: 'There is a big difference between answering students' questions about creationism and actually introducing it into the lessons in the first place as part of the curriculum. If the teacher raises the topic, then it takes on an authority that it does not deserve. Hampshire should think again about this proposal. It will add nothing to the education of children, but will create confusion in their minds about what is science and what is religion'.

The UK Government decision against the Truth in Science distribution does not block attempts to introduce Intelligent Design teaching into Scottish schools. Scotland sets its own standards in education. Her Majesty's Inspectorate of Education (HMIE) is an executive agency of the Scottish Government, with 'responsibilities to evaluate the quality of preschool education, all schools, teacher education, community learning and development, further education and local authorities'. Qualifications at the secondary school level and for further education are provided by the Scottish Qualifications Authority. The following is quoted from an article in the *Sunday Herald*, a Scottish newspaper, from June 2007, following the UK Government announcement:

> INTELLIGENT DESIGN, a controversial alternative theory to evolution, could become part of the science curriculum in Scottish schools [...] the Scottish Qualifications Authority (SQA) is considering provision for the theory as part of a review of the science course curriculum [...] Intelligent design (ID) is one of a wide range of theories of origin currently taught as part of the Religious, Moral and Philosophy Studies (RMPS) SQA course, but could be moved elsewhere as part of the review. A spokesman for the SQA said: 'It happens to sit in RMPS just now. If and when it does become part of the curriculum for science, which it may well do as part of this review, then that's where it could sit'. [...]
>
> [The] director of education for Inverclyde is not in favour of prohibiting Truth in Science material and accepts teachers are free to present ID informally. He said: 'I have no objection to intelligent design being advanced as one theory, but most teachers don't have time. I trust head teachers to make their own decisions about what is appropriate'. [...]
>
> An education spokeswoman for the Scottish Executive said: 'We're not prescriptive as to books or materials. We provide guidelines, and within those guidelines it's up to schools to decide'.

Considering the evidence, one might hazard a guess what would be the outcome, if the teaching at our local primary school is representative of Scotland and the UK at large, thereby raising a question over the prevalence of Creationism in our education system, as a whole. But, if it does, then it also raises one about the teaching of evolution, on the whole. It's very clear that personal religious beliefs can get in the way of presenting science honestly. The extreme consequence of this is that science is not taught at all.

Richard Dawkins' 2006 two-part series *The Root of all Evil*[8] visited the Phoenix Academy in London, an evangelical school adopting the American Baptist Accelerated Christian Education curriculum that integrates Christianity with educational subjects, such as Noah's Ark in science lessons. Adrian Hawkes is the head teacher at the school:

> Running a Christian school within the independent sector often means that we are contacted by TV to do this or that, the often asked question is about evolution and creationism as if somehow we did nothing else. Even people like Richard Dawkins have spent time grilling us on our views for his documentary *The Root of all Evil*. Funnily enough we don't usually think much about Darwin or evolution between those TV times.

From what I know of Darwin he was a good family man, doing interesting research; however, what is very important to me is what people really think. I think that Darwin himself was still in the realm of theory as to how and why. Today we tend to be more dogmatic and sure as per Richard Dawkins, when maybe we ought not to be so proud of our certainties.

Our thinking is that, if the Bible is correct (and I believe it to be so), then our thoughts are the things that mould us and make us (*as we think in our hearts so we are* – Proverbs 23:7), for our thinking is that which generates the way we act. So, if we think that the universe is some mechanical process, then we tend to treat people like a machine; if we think that we are just an animal, then we tend to treat each other like animals.

It seems to me that Hitler believed that Darwin's view of how life worked was correct and much more than theory. From this belief he decided (thought), 'Well then, let me speed up the process and create an evolutionary jump and make the master race'. We know the results!

If, on the other hand, we believe (think) that there is a mind behind our universe, and that we are created in the image of God then surely that thinking should encourage us to treat one another with dignity and respect! Richard Dawkins' argument with me on TV was that he was more honourable than me as he did not need a God to stop him from pillaging and raping; his implication being that I did! My problem with that argument is that one person cannot negate what is going on in the world with great swathes of destruction and man's inhumanity to man. Our thinking that we can play God, and our devaluation of each other, is a daily fact whatever Richard thinks.

So what do I think of Darwin? Interesting theories; but believing them to be so totally correct without criticism leads to dangerous thought.

The UK surveys and censuses show us that there is little consensus on Darwinism: all of those segments of society grouped by profession – biologists, teachers, politicians, and everyone else – all of those groups exhibit some dissent over evolution. And yet, most will have been educated by the same system, which suggests that there may be difficulties in getting the message across. So, when faced with the responsibility to teach Darwinism honestly, what are the challenges? Geoff Morgan is a biology teacher at a private school in Edinburgh:

From a secondary school teaching point of view, one of the problems is that the concept of evolution by natural selection is so simple. It's a very simple idea, but the ramifications of that simple idea are very complex for a pupil to take on board. A pupil can learn to parrot the theory without understanding its all-encompassing power, its overarching importance to life on the planet. You will get students who will be able to score good marks in tests on that topic but who couldn't explain biodiversity on the planet. The phrase 'survival of the fittest' shoots the bolt, once you've heard that there's no punchline!

And this would seem to be the enduring problem with Darwinism: it's such a simple idea, but it has grand repercussions. The distance between meaning and mechanism transects everything that we are. In Richard Dawkins' words, it has a *colossal ratio*:

> Darwin is so important, it is almost absurd that children don't learn about it when they are tiny, practically because it is the explanation for our existence, the existence of all living things, and it's in a way, one of the most powerful explanations of anything that anybody has ever suggested, because if you think about it, the ratio of that which it explains, which is everything about life and everything about complexity divided by that which you need to postulate in order to do the explaining, is a colossal ratio, because what you have to postulate is extraordinarily simple. It actually amounts to little more than high fidelity genetics because everything else kind of follows naturally. And from that almost miniscule level of assumptions you can explain just about everything about life, and yet, simple as the explanation is, powerful as it is, and huge as the magnitude of what it explains is, nobody thought of it before the middle of the 19th century. Which is an astonishing fact because it doesn't require great mathematics, doesn't require highly sophisticated notation of any sort, anybody can understand, although an amazing number of people fail to. So, it's almost unique, perhaps it is unique, in the sheer power of what it does for the human intellect.

A biology teacher in Edinburgh said that the difficulty with teaching Darwin is it has 'no punchline'. It's simply a very quick statement of a mechanism and then there's no corollary.

Well, in a way the punchline is 'everything'.

So, at what age do you start teaching Darwinism?

I'd like to give it a try at about 8, but I haven't tried. I hope to write a book for something like 10-year-olds.

Epilogue

I have tried to maintain a fairly non-partisan voice throughout this book, to give my interviewees fair hearing while presenting their disparate opinions. The historical sections should have been descriptive and factual; the interviews, probing and informative, without my narrative biasing your reading of them. However, perhaps here I can allow myself to make some general comments of my own.

Writing this book, I inevitably had to get drawn into the religion versus science debate; it never seems far from Darwinism's door. Even while structuring it, the raw material fell neatly divided; but are those subdivisions equally balanced? It's certainly surprising to discover that Creationism, which isn't really renowned for its grounding in biology, actually has more acceptance of Darwinian ideas than Intelligent Design, which does claim biological authority. Nonetheless, purely from the contents herein, and trying hard to not apply any personal bias, it seems that the evolutionary scientists quoted here at least appear to profess a deeper, mechanistic understanding of their subject than their challengers, and while the debate involves us all, this battle is, after all, being fought within the scientific arena. Technical jargon aside, the scientists' understanding of the relevant processes clearly outdoes the woefully small sample that I have collected of alternative viewpoints. But, even looking beyond the covers of this book, the science can't all be rhetoric: there is a tough regime of exactitude that oversees it. And then there's the mass of evidence in its favour, and I really can't help but wonder where is the equivalent contradictory evidence?

To me, 'How do you know that it's not the word of God? Were you there?' has as much use for argument as it's antithetical, 'How do you know that it is the word of God? Were you there?', and when such a stalemate occurs, the only recourse is to look at the evidence. Even if you are not an expert in evolutionary biology, be assured that thousands of independent sources all pointing towards the same logical conclusion simply cannot be dismissed as the biggest conspiracy hoax in history. Far from there being a controversy, I find the evidence for Darwinian evolution undeniably incontrovertible. In contrast, historical hearsay and claims about a book's Divine source don't prove anything, partly because of the usually cited lack of testability, but mainly because the evidence in support of it having a supernatural source is so weak. So, while I have enjoyed, and have been very grateful of, conversations with the Creationists and proponents of Intelligent Design who have kindly contributed to this book, I have to admit that, if I did want to confirm a detail of nature, I think I would rather seek scientific authority.

Science is an approach to the world that I recognise in Darwin's extensive and wonderful works. The discernment of truth seems so far better founded in tangibles, through experimentation and observation, than untested, ethereal explanations for life. This rationalism is the attitude that Darwin absorbed from his time in Scotland, and particularly at Edinburgh; from Hume's empiricism that illuminates almost every thought behind Darwin's work, through to Edmonstone who played a vital role in giving Darwin ideas of travel, perhaps a new-found maturity, and the means to amass the body of evidence that contributed to his conclusions. His time with Grant was also pivotal, in his regard of alternative paradigms in natural history, and the experience of examining nature, and forming opinion on it. Exactness in presenting his findings and constructing clear argument came from his extracurricular time spent with the revolutionarily anarchic Plinians. Later in life, he received support and direction from his Scottish friend Hooker, and also inadvertent assistance from Jenkin. And then there's his reading of Scottish authors, particularly Chambers and Lyell, and therefore, by association, Hutton as well. Really, his most valuable lessons were learnt beyond any institution. Perhaps equally important were the repulsive influences in his life. The detestation of medicine. The boredom of lectures. The elements that 'wasted' his education.

Perhaps the essence of this book is education. Education was an important factor paving the way to the Age of Enlightenment, and was particularly noticeable in the role it played in the Scottish Enlightenment; an educated populace was amenable to new understandings in science. The scientific content of education was an area of concern even in Darwin's day, and even before publication of the *Origin of Species*. At that 1855 British Association meeting in Glasgow, its president, the Duke of Argyll, whom Darwin thought spoke excellently, emphasised the need for 'securing for [science] a better and more acknowledged place in the education of the young'.

One hundred and fifty-four years later, an excellent debate organised by the Institute of Ideas and the Humanist Society Scotland was held in the National Library of Scotland on Darwin's bicentenary eve. Most of the usual viewpoints were represented and the debate largely revolved around the teaching of Darwinism and Intelligent Design in UK schools. The Old Earth Creationist argued that questions about the Big Bang and the multiverse are beyond science and necessitate a move into the realms of philosophy. Unsurprisingly, the Intelligent Design supporter picked holes in Darwinism, to expose its incompleteness. The two pro-science writers called for Creationism to be taught in schools, but not in science classes. The secondary school science teacher thought that the reason that everything from Lamarckism to Creationism makes its way into science lessons in the first place is because we are obsessed with teaching the history of science, particularly the critiquing of the making of science: in his opinion, this obsession is a secular problem, and an unfair expectation of the student, and, he implored, 'Why don't we just teach some science?'

I would go further: the modern debate has missed the point, but only by a hair's breadth. It does mostly conclude that science and Creationism should be taught separately in schools, and that, of the two, only Darwinism should be taught in science lessons. But, core to that, it should be explicitly separated from anything abiotic and irrelevant. Darwinism starts with the living. It has nothing to say about abiogenesis, Creation, Big Bangs, or anything unrelated to Darwinian evolution. Be assured that it was not an evolutionary biologist who forged the link between the origin of life and Darwinian evolution. Darwin never claimed it, in his professional nor personal life. In the *Origin of Species*, he solely contradicted the immutability of species in Natural Theology. He was most realistic about the limits to his ideas: 'some naturalists [...] believe that many structures have been created for the sake of beauty, to delight man or the Creator (but this latter point is beyond the scope of scientific discussion)'.

Darwinism was likely dragged into the various timelines of Creationist interpretations, which do deal with beginnings, because of the corollary that follows on from evolution by natural selection, that a deity was not required for evolution of life. Darwin simply explained how all the 'elaborately constructed forms, so different from each other, and dependent upon each other in so complex a manner' that we see could have come about. His Huxley-style agnosticism purely put him in a freethinking position from where to observe without bias and without preconception. In retrospect, it seems so unfair that Darwin and his work has been portrayed as anything but a search for evident truth based upon David Hume's empirical values. Instead, Darwinism has reached a duality: it is both the Anti-Christ and the main rationale for atheism. Darwin is being lambasted with no right of reply, and it seems unfair that his work is claimed to stand for something unrelated, or when depictions of him take on the guise of the devil. And so it is very notable, especially considering the attention the human eye receives from anti-Darwinists, and Darwin's own agnosticism, that in the *Origin of Species* he expressly measures the efficacy of his natural selection against the handiwork of a Creator, whom he clearly considered to be beyond the measure of man:

It is scarcely possible to avoid comparing the eye to a telescope. We know that this instrument has been perfected by the long-continued efforts of the highest human intellects; and we naturally infer that the eye has been formed by a somewhat analogous process. But may not this inference be presumptuous? Have we any right to assume that the Creator works by intellectual powers like those of man? If we must compare the eye to an optical instrument, we ought in imagination to take a thick layer of transparent tissue, with a nerve sensitive to light beneath, and then suppose every part of this layer to be continually changing slowly in density, so as to separate into layers of different densities and thicknesses, placed at different distances from each other, and with the surfaces of each layer slowly changing in form.

Further we must suppose that there is a power always intently watching each slight accidental alteration in the transparent layers; and carefully selecting each alteration which, under varied circumstances, may in any way, or in any degree, tend to produce a distincter image. We must suppose each new state of the instrument to be multiplied by the million; and each to be preserved till a better be produced, and then the old ones to be destroyed. In living bodies, variation will cause the slight alterations, generation will multiply them almost infinitely, and natural selection will pick out with unerring skill each improvement.

Let this process go on for millions on millions of years; and during each year on millions of individuals of many kinds; and may we not believe that a living optical instrument might thus be formed as superior to one of glass, as the works of the Creator are to those of man?

Appendix: Clarification of Terms

Communicating the ideas that form Darwinism seems to have suffered more than most from the evils of paraphrasing and misquotation. For Darwinian evolutionary theory, particularly natural selection and gradualistic speciation, the misconception of these ideas has arguably led to misappropriation of its underlying mechanisms in the search for analogous processes. Much misunderstanding has arisen out of the expressions that have been coined by others to describe those processes. Thus, instead of forwarding an understanding of evolution, such expressions have done greater damage than if Darwin's words had been left to speak alone. This practice first came fast on the heels of the publication of his greatest work, the *Origin of Species*. Catch phrases were coined, passages paraphrased and disciplines devised. The struggle to make natural selection fit has been going on for 150 years. 'Survival of the fittest'? The phrase itself is an anathema to Darwinists. It and several more of the many examples of imprudent acts by use of impudent diction that have caused damage to Darwin's name are collected together here by John Llewelyn[1] towards a clearer understanding of Darwinism:

> The anniversary of the publication of the *Origin of Species* is a good time to remind ourselves how some of the expressions employed in discussions of it have been misread, sometimes in ways that have given rise either to misplaced pessimism or to misplaced optimism.

Survival of the fittest

> This phrase is due to Herbert Spencer who outlined an evolutionary philosophy a year before the publication of the *Origin of Species* appeared. The phrase caught on as a handy means of getting across the gist of Darwin's scientific hypothesis. However, in so far as it is applicable to Darwin's biological hypothesis it must be understood as referring not to particular members of a population, but to the group as a whole. The 'fittest' will be that group which increases its numbers more rapidly than other groups in the same geographical environment. To say that the group survives is to add very little to what is said when it is asserted that it is the fittest. To say that it is the fittest is to say that it survives because

members of the group are so constituted biologically that they make better use of the food, mating and other resources available in the particular geographical situation under consideration. To say that a group is the fittest is not to say that its members are, so to speak, in better athletic form than the members of all other groups. A particular member of a group can be fit in the Darwinian sense of prolific or able to propagate and at the same time unfit in the sense of ailing or weak.

Spencer's phrase 'survival of the fittest' is Darwinian only when read as equivalent to 'survival of the best fitted or best adapted'.[2]

Struggle for existence

This phrase has been taken literally and has led T.H. Huxley and others to the mistaken belief that Darwin provided a scientific basis for the view that nature is 'red in tooth and claw' [from 'In Memoriam A.H.H.' by Alfred, Lord Tennyson, 1849], that organisms are naturally bellicose and that life is one long (or short) all-in wrestling match. Because he believed this, T.H. Huxley, unlike his grandson, of whom further mention will be made below, argued that it was man's duty to combat evolution.

However, Darwin himself uses the expression 'struggle for existence' as a metaphor for the manner in which one species persists in a certain environment, whereas others disappear. But the persistence of the one species follows from the fact that its members happen to be better equipped constitutionally for life in that locale than are members of other species. This does not mean that they are better fighters than the others. It is easy to imagine situations in which pugnacity and gladiatorial prowess would be less conducive to survival than timidity or quietism.

T.H. Huxley's pessimism is grounded on a misconception, a misconception, incidentally, that is similar to popular misreadings of the phrase 'will to power' as used by that anti-Darwinian Friedrich Nietzsche.

Origin of species

Darwin's title has led to pessimistic inferences by Christians who, not having read the book, supposed that it had to do with the Creation of life, in particular human life. Darwin does use the word 'Creation' in the book, for example in the concluding paragraphs, but later, in view of the misinterpretations this gave rise to, expresses the wish that he had not used that particular term. The main concern of the scientific hypothesis proposed in the *Origin of Species* is the effects of natural selection in the light of variations. It is not concerned with the origin of variations, of which Darwin tentatively accepted the explanation given by Lamarck in terms of the inheritance of acquired characteristics. Nor is it concerned with the question of the origin of life, which, Darwin writes in one of his letters, is not a question for science.

True, many would still consider the prospect opened up by Darwin to be dismal because, as they put it, Darwin explains the higher in terms of the lower, deriving the human species from an origin that is below the dignity of humanity. Darwin's own response to this is to say that he would prefer to be descended from an ape that risks its life to save its keeper than from a savage who relishes torturing his enemy, treats his wives as slaves and is himself a slave to monstrous superstitions.

Natural selection

Some of Darwin's expressions misled the reader into supposing that there is a Grand Selection Committee sitting around a table somewhere with the agenda of picking a First XV and a brace of reserves. Darwin's theory, however, implies not that there is an independent authority that does the selecting, but that the selecting goes on within the group of candidates for places in the team. Hence, those critics who say that it is nonsense to teach, as Darwin does, that there are blind powers directing the course of evolution are barking up the wrong tree. They are mistaken on at least two points, first in assuming that Darwin teaches that the power which does the selecting is blind and mechanical, second in assuming that the only alternative is a power that does the selecting with vigilance and with purpose. There is a further alternative, the one adopted by Darwin. For him the power that does the selecting is neither mechanical nor purposeful, because there is no power that does the selecting.

Darwin himself is not entirely free from blame for these misconceptions. The terms 'selection', 'fitness', 'struggle for existence', 'survival of the fittest', which he introduces or takes over from others to expound his theory, all too easily suggest that there are ends aimed at, whether consciously or unconsciously or otherwise. In *The Descent of Man* he confesses that in the earlier editions of the *Origin of Species*, although he had considered that he was supplanting Lamarck's teleological theory with a non-teleological one, he himself had not been able to exclude teleological ways of speaking as thoroughly as he wished. He admits:

> I was not ... able to annul the influence of my former belief, then almost universal, that each species had been purposely created; and this led to my tacit assumption that every detail of structure, excepting rudiments, was of some special, though unrecognized, service ... hence if I have erred in giving to natural selection great power, which I am very far from admitting, or in having exaggerated its power, which is in itself probable, I have at least, as I hope, done good service in aiding to overthrow the dogma of separate Creations.[3]

Darwin's use of the language of purpose reflects the general difficulty of talking about living beings without it. Teleological metaphors and analogies can assist in the exposition of non-teleological ideas. Kant argues in the *Critique of Judgement* that teleological tropes are unavoidable, yet no one insists more strongly than he that to take teleological terms literally is unscientific.

Law of evolution

Karl Popper has argued that it is doubtful whether the phrase 'law of evolution' denotes a topic in the scientific theory of Darwin unless it is intended as a compendious way of referring to the set of laws according to which species succeed one another, for instance laws of natural selection, sexual selection and heredity. Consider the first of these, which C.H. Waddington formulates as follows: 'If more offspring are conceived than are necessary to preserve the numbers of the population, and there are inherited variations between individuals, … then some variant types that are weaker than their fellows will more often die before becoming parents, while others, the more efficient ones, will more frequently survive to pass their qualities on to the next generation'.[4] This is a testable law because the conditions referred to in the 'if' clause can be repeatedly found, and it will be possible in each case to check the validity of the prediction made in the 'then' clause.

This scientific sense of the phrase 'law of evolution', however, is quite different from the sense it has been given by certain philosophers of history and certain social theorists who employ the phrase of what they conceive as a single process determining, in the words of T.H. Huxley, 'the unvarying order of that great chain of causes and effects of which all organic forms, ancient and modern, are all links'.[5] The origin of this idea predates Darwin, but to some of his readers it seemed to be put on a scientific basis by him. Oswald Spengler, Arnold Toynbee and some Marxists applied what they took to be Darwin's scientific law to society and history. But, as H.A.L. Fisher [1865–1940] writes, although 'Men wiser and more learned than I have discerned in history a plot, a rhythm, a predetermined pattern … I can see only one emergency following upon another…, only one great fact with respect to which, since it is unique, there can be no generalizations…'.[6] Therefore, if we are inclined toward optimism or toward pessimism by reflection on the laws of social development that Comte, Marx, Spencer or Spengler claim to have discovered, we should reflect too on the fact that in spite of the phrases these writers borrow from science, their so-called laws of evolution are not certified scientifically, and, in particular, not certified by or in the manner of Darwin's biological law of natural selection.

The point is not that there can be no scientific laws of society. Darwin's preceptor Malthus is witness to the untenability of such a denial. The point is rather that some philosophers and social theorists have used the expression 'law of evolution' in an unscientific way while trading on the circumstance that Darwin gave it a quite scientific use in biology.

Scientists too, even biologists, have attempted to extend the application of Darwin's hypothesis beyond the field of the biological, expecting it to do work which, for reasons of logic, it cannot do. The culprits include Waddington and Julian Huxley. Both of these have attempted to show how the theory of evolution has put ethics on a scientific footing. Why are their attempts misconceived?

Evolution and ethics

In his Romanes Lectures Julian Huxley comments on his grandfather's Romanes Lectures as follows:

> For T.H. Huxley, fifty years ago, there was a fundamental contradiction between the ethical process and the cosmic process. By the former, he meant the universalist ethics of the Victorian enlightenment, bred by nineteenth-century humanitarianism out of traditional Christian ethics.... And the cosmic process he restricted almost entirely to biological evolution and to the struggle for existence on which it depends. 'The ethical progress of society' – this was the main conclusion of his Romanes lecture – 'consists, not in imitating the cosmic process, still less in running away from it, but in combating it'.

> Today that contradiction can, I believe, be resolved on the one hand by extending the concept of evolution both backward into the inorganic and forward into the human domain, and on the other by considering ethics not as a body of fixed principles, but as a by-product of evolution, and itself evolving.[7]

Are we uncomfortable with Julian Huxley's and Waddington's attempts to trace our ideas of ethical right and wrong to a non-moral source because we are of the opinion that if it is possible to give the *cause* of our believing that such-and-such then no justifying *reason* can be given why we should hold that belief? This opinion is mistaken. Thus an anthropological description of the origin and history of our ethical code may indeed weaken any commitment we may have made to that code. It is psychologically difficult to maintain simultaneously the attitude of committed participation in a community defined by subscription to that code and the attitude of an external observer. This psychological difficulty of maintaining a position that is on the threshold between an insider's engaged endorsement of a prescription and an outsider's report on the prescription is to be distinguished from an alleged logical inconsistency between genetic explanation and rational justification. But although the proponents of evolutionary ethics mentioned above may not be culpable of failing to make this distinction, they remain vulnerable to criticism in another respect.

The biological theory of natural selection is a scientific analogue of the liberal picture of society and of John Stuart Mill's faith that truth will out if only everybody is allowed to have his or her say and we are allowed to talk for long enough. But how long is long enough? Pragmatists like Charles Sanders Peirce, who owes a lot to Darwin, goes as far as to say that 'The opinion which is fated to be ultimately agreed to by all who investigate, is what we mean by the truth, and the object represented in this opinion is the real'.[8] But, again, how do we know when we have reached the ultimate? Don't we have to wait until the opinion on which everyone agrees is a *true* opinion? Similarly when it is said that evolution points the way to ethical truths and principles that we ought to adopt. Even if we could agree that human evolution makes a bee-line for the best – a proposition that may have seemed more reasonable to Victorian Englishmen than it seems to us today – we want to know how long we have to wait. Julian Huxley admits that the selection of ethical principles is not a natural selection and that human decisions play the main role in determining which are to be accepted as criteria of how we ought to behave – which sounds suspiciously like admitting a tautology. So at any given moment it is open to a person to advance the course of the evolution of moral ideas by opting for this or that as a criterion of what is the right thing to do. You can, for example, decide to accept the criteria or principles that are generally accepted in this or that society, or you can adopt novel criteria, depending on whether you think this or that society has or has not reached the apex in the evolution of its moral ideas. If you think that there is still advance to be made and that it is open to one to 'lessen the birth pangs' of a superior moral outlook, you will be faced with a choice among a multiplicity of novel criteria any one of which may later catch on and become a further stage in the history of the improvement of moral principles. But all these decisions presuppose that you have made up your mind as to just what is most worthwhile. However accurate may be your judgement about what moral principles or criteria people have accepted, do accept or will accept, you cannot infer from this judgement that it is the principle or criterion which you ought to accept. The history or evolution of moral ideas can help you only if you have decided in advance that this history is tending in the direction of improvement. But that it is tending in the direction of improvement is a verdict one can pass only if an answer has been given to the question what things are worth having. This question is not an empirical one. Although one of the reasons why a belief or principle survives may be that it is true or right, beliefs are not true and principles are not right because they survive.

Advocates of evolutionary ethics like Julian Huxley and Waddington may respond to these comments that although evolution is of no use as a guide to what things we should judge to be valuable for their own sakes, evolution can teach us what actions are the most likely to lead us to these things. But does this mean anything more than that knowledge or experience helps us to see how we can bring about what we want to bring about? To learn this there was no need for us to wait for Charles Darwin.

In short, to state that the course of evolution can tell us what we ought to do is either to state something that is true because vacuously tautological, or it is to state something that is empirically and scientifically false.

Postscript

The dilemma that has just presented itself will arise if we do not make clear to ourselves whether our idea of the evolved is or is not already ethically evaluative. It raises the question whether an evolution of the idea of evolution is conceivable that moves beyond a clear-cut antithetical distinction between the logical and the empirical, for instance the psychologically or biologically empirical, and beyond a sharp separation of logically analytic from genetic and historical discourse. Anyone who shares my interest in this question will need to read another classic the anniversary of the appearance of which is still being celebrated, Hegel's *Phenomenology of Spirit*. There, as in other works of his, history is interpreted by Hegel as a series of negations and contra-dictions, therefore as a logical process. Called for too by anyone intrigued by the possibility of synthetic apriority will be study of the genetic phenomenology of Edmund Husserl and his successors.

Meanwhile, our interest in evolution in a more Darwinian empirical sense has been revived by a project recently launched by the Ian Ramsey Centre for Science and Religion at Oxford. Having gone to some pains in a recent book to challenge the persistent assumption that the notion of the religious must be defined in terms of the religions (an assumption made, it seems to me, by Richard Dawkins in *The God Delusion*), I look forward to learning that participants in the Oxford project will not confuse the application of evolutionary theory to the idea of God and its application to the idea of religion.[9]

Endnotes

CHAPTER 1

1 The following historical account of the building of the University of Edinburgh is adapted from Andrew G. Fraser's presentation 'The University Natural History Museum' at the *Darwin's Edinburgh* symposium, 14 March 2008.

2 'Edinburgh University's strict graduation requirements meant that candidates had to have studied medicine for at least three years, one year of which had to be at Edinburgh itself. They had to have attended lectures in all the courses offered by the medical faculty except Midwifery, though it too was recommended. Finally, they had to undergo a series of oral and written examinations in Latin. They also had to compose a Latin thesis and defend it before the whole Faculty. Despite the use of Latin, the examination did not test knowledge of classical medical literature as it did at some Continental universities. Instead the traditional literary form of the medical doctorate was transformed into a thorough test of the candidate's knowledge of the medical subjects taught in lectures' (Rosner 1992).

CHAPTER 2

1 Aristotle developed his theory of inheritance in *Generation of Animals* and *Parts of Animals*, both completed *c.*350 BC. However, these works were predated by the extensive tutorial *Physicae Auscultationes* which prepared the way for his subsequent biological works. In it, Aristotle is essentially posing the question 'Why?' by defining Four Causes underlying life. He divides the 'nature' of everything biological into its 'material nature' and its 'formal nature' (Henry 2006). The latter is then subdivided into nature 'as end', the adult form, and nature 'as mover', the essence of the adult within the embryo; a remarkable insight into inheritance arrived at from his experiments: he discovered that even if parts of the parent plant are removed before seeds are produced, those missing parts still appear in the seedling, ancient empirical proof that what is directly transmitted in the act of reproduction is not the observable part of an organism (the phenotype) but its underlying cause (the genotype). Aristotle concluded that 'propagation implies a creative seed endowed with certain formative properties [...] the parent animal pre-exists, not only in idea, but actually in time. For man is generated from man; and thus it is the possession of certain characters by the parent that determines the development of like characters in the child'.

2 Aristotle taught abiogenesis or spontaneous generation (otherwise, *Generatio aequivoca*, *Generatio primaria*, archegenesis, autogenesis, and archebiosis) based on his 'observations' that fleas and some other animals arise out of putrid matter and aphids from dew on plants, whereas mice come from dirty hay and crocodiles from rotting logs on river beds. He was largely speculating on the unobservable, quickly refuted by scientific study: Francesco Redi proved in 1668 that maggots did not appear in meat when flies were obstructed from laying their eggs by wire screens, concurrent with the advent of microscopy, laying the ground for *omne vivum ex ovo*, life from the living.

3 The student intake figures of Ashworth (1935) and the population estimates from a few years earlier suggest about a 2.4% influx for 1825–26, compared with 5.1% today.

4 See Rosner (1992).

5 Scotland & Medicine: Collections and Connections: http://www.scotlandandmedicine.com/site/CMD=PICDETAIL/PICID=248/838/default.aspx

6 See Keegan (1996).

CHAPTER 3

1 See Chapter 2, n. 1.

2 Here it is Aubrey: p. 36 and Table 9 in *The Different Forms of Flowers on Plants of the Same Species*. This book was an extensive reworking of an earlier paper from 1862, 'On the two forms, or dimorphic condition, in the species of *Primula*, and on their remarkable sexual relations', published in the *Journal of the Proceedings of the Linnean Society of London (Botany)*.

3 Extracts from *The Different Forms of Flowers*: 'It has long been known to botanists that the common Cowslip (*Primula veris*, Brit. Flora, var. *officinalis*, Lin.) exists under two forms, about equally numerous, which obviously differ from each other in the length of their pistils and stamens. This difference has hitherto been looked at as a case of mere variability, but this view, as we shall presently see, is far from the true one. Florists who cultivate the Polyanthus and Auricula have long been aware of the two kinds of flowers, and they call the plants which display the globular stigma at the mouth of the corolla, "pin-headed" or "pin-eyed", and those which display the anthers, "thrum-eyed" [in Johnson's Dictionary, thrum is said to be the ends of weavers' threads; and I suppose that some weaver who cultivated the polyanthus invented this name, from being struck with some degree of resemblance between the cluster of anthers in the mouth of the corolla and the ends of his threads]. I will designate the two forms as the long-styled and short-styled [...] The results of my trials on the fertility of the two forms, when legitimately and illegitimately fertilised, are given in Table 9 [...] Twenty-three spontaneously self-fertilised capsules from this form contained, on an average, 19·2 seeds. The short-styled plants produced fewer spontaneously self-fertilised capsules, and fourteen of them contained only 6·2 seeds per capsule'.

4 The passage in full from the first edition: 'It is interesting to contemplate an entangled bank, clothed with many plants of many kinds, with birds singing on the bushes, with various insects flitting about, and with worms crawling through the damp earth, and to reflect that these elaborately constructed forms, so different from each other, and dependent on each other in so complex a manner, have all been produced by laws acting around us. These laws, taken in the largest sense, being Growth with Reproduction; Inheritance which is almost implied by reproduction; Variability from the indirect and direct action of the external conditions of life, and from use and disuse; a Ratio of Increase so high as to lead to a Struggle for Life, and as a consequence to Natural Selection, entailing Divergence of Character and the Extinction of less-improved forms. Thus, from the war of nature, from famine and death, the most exalted object which we are capable of conceiving, namely, the production of the higher animals, directly follows. There is grandeur in this view of life, with its several powers, having been originally breathed into a few forms or into one; and that, whilst this planet has gone cycling on according to the fixed law of gravity, from so simple a beginning endless forms most beautiful and most wonderful have been, and are being, evolved'.

5 From the final paragraph: 'There is grandeur in this view of life, with its several powers, having been originally breathed *by the Creator* into a few forms or into one; and that, whilst this planet has gone cycling on according to the fixed law of gravity, from so simple a beginning endless forms most beautiful and most wonderful have been, and are being, evolved'. The italicised phrase [my italics] was introduced after the first edition.

6 Anne Elizabeth 'Annie' Darwin (2 March 1841–23 April 1851), second child and eldest daughter of Charles and Emma Darwin, fell ill in 1849 having caught scarlet fever at the same time as her two sisters, and perhaps also tuberculosis.

7 This abstract from Grammer *et al.* (2004): 'The relationship between a female's clothing choice, sexual motivation, hormone levels, and partnership status (single or not single, partner present or not present) was analyzed in 351 females attending Austrian discotheques. We digitally analyzed clothing choice to determine the amount of skin display, sheerness, and clothing tightness. Participants self-reported sexual motivation, and we assessed estradiol and testosterone levels through saliva sampling. Results show that females are aware of the social signal function of their clothing and that they in some cases alter their clothing style to match their courtship motivation. In particular, sheer clothing – although rare in the study – positively correlated with the motivation for sex. Hormone levels influenced clothing choice in many groups, with testosterone levels correlating positively with physique display. In females who had a partner but were at the disco unaccompanied by the partner, estradiol levels correlated positively with skin display and clothing tightness. Significant differences were not found, however, for clothing choice across the partnership-status groups'.

CHAPTER 4

1 Darwin's family moved to Down House in September 1842. The first record of his having ideas of common descent that would shape the *Origin of Species* appears on a page headed by 'I think' then carrying a roughly sketched evolutionary tree in the famous single greatest advancement of Darwin's ideas was not made until just prior to moving in May; he attempted to formalise his thoughts by writing a short essay, *Sketch on Natural Selection*. Subsequent work was produced in Down House or during short excursions to seek remedial treatments.

2 He used both hands, silly!

3 The ambiguity here is because there is some evidence that Jameson was an equal to Grant in his Lamarckism, despite his biblical leanings, authoring that first notable paper himself (see Secord 1991).

4 See Avery (2003).

5 Available at http://darwin-online.org.uk/EmmaDiaries.html

6 See Musical Times (1901).

7 This is a young and burgeoning research area, but if you want to read further, try these: Steele (2008), Spadafora (2008), Marques *et al.* (2005) and Honeywill (2008), as well as the less technical Derry (2009d) and Lawton (2009).

8 He also pointed out that while natural selection could bring about adaptation, the *Origin of Species* doesn't actually discuss species, nor their origins.

9 See Coyne and Orr (2004).

10 Inspired by Wright, Gustave Malécot's doctorate on the correlation between relatives eventually led him to discover this fundamental idea which he published between 1941 and 1948, through application to inbreeding, the correlation between relatives and random mating in a finite population. In 1940, Charles Cotterman had independently made essentially the same discovery during doctorate study, which he went on to use in many aspects of human genetics.

11 See Holland (1975).

CHAPTER 5

1 Jameson's 1826 course in zoology concluded with lectures on the philosophy of zoology, starting with *Origin of the Species of Animals* (Secord 1991).

2 Although Croll enjoyed the environs of the city, and the countryside near his home in Morningside, he reported to Darwin a decline in its academia: 'Edinburgh, with all its books and learning, is miserably behind in scientific literature. Since I came here, I hardly know what is going on in the scientific world around. One can get plenty of good solid books on science, but the current news and literature of the subject are not to be found anywhere. Edinburgh, I fear, is falling behind' (as recorded in Irons 1896).

CHAPTER 6

1 Robert Louis Stevenson's (1907) biography of Jenkin hardly mentions this episode, but turn to Jenkin (1867) for the original, and Bulmer (2004) for a recent commentary.

2 It has been long thought that Darwin was dramatically compromised by Jenkin (e.g. see Willis 1940).

3 See Gould (1985a).

4 Asked in 1895, when testifying as a defence witness during his trial for gross indecency, to define what was meant by this phrase in Lord Alfred Bruce Douglas' poem 'Two Loves', Wilde explained: '"The love that dare not speak its name" in this century is such a great affection of an elder for a younger man as there was between David and Jonathan, such as Plato made the very basis of his philosophy, and such as you find in the sonnets of Michelangelo and Shakespeare. It is that deep spiritual affection that is as pure as it is perfect. It dictates and pervades great works of art, like those of Shakespeare and Michelangelo, and those two letters of mine, such as they are. It is in this century misunderstood, so much misunderstood that it may be described as "the love that dare not speak its name", and on that account of it I am placed where I am now. It is beautiful, it is fine, it is the noblest form of affection. There is nothing unnatural about it. It is intellectual, and it repeatedly exists between an older and a younger man, when the older man has intellect, and the younger man has all the joy, hope and glamour of life before him. That it should be so, the world does not understand. The world mocks at it, and sometimes puts one in the pillory for it.' The courtroom transcript reports that this elicited 'Loud applause, mingled with some hisses'.

5 Materialism stemmed from the philosophies of Thales, Aristotle and Lucretius, amongst others, and simply holds that the only thing that can be proven to exist is matter itself: all things are composed from matter, and phenomena result from their material interactions. Aristotle's material cause (or nature; see Chapter 2, n. 1) simply describes the material out of which something is composed, while the epic poem by Lucretius, 'De rerum natura' ('On the Nature of Things'), speaks of matter comprising assemblies of 'atoms' and depicts the principles that 'nothing can come from nothing' and 'nothing can touch body but body'. The Scottish Enlightenment extended this understanding. In particular, Hume further defined materialism in the context of rational thought, in his An Enquiry Concerning Human Understanding (Hume 1748): 'Is there any principle more mysterious in all nature than the union of soul with body; by which a supposed spiritual substance acquires such an influence over a material one, that the most refined thought is able to actuate the grossest matter? Were we empowered, by a secret wish, to remove mountains, or control the planets in their orbit; this extensive authority would not be more extraordinary, nor more beyond our comprehension ... The only method of undeceiving us is to mount up higher: to examine the narrow extent of science when applied to material causes'. Anti-materialism therefore posits an opposing view, that matter is not the only substance, hence the 'soul' in the case of humans.

6 See Darwin (1887), but more complete as Darwin (1958).

7 Other than his revulsion for Darwinism, and his sometimes plagiaristic popularisation of scientific ideas, including aspects of glacial theory, Agassiz was also a racist who supported apartheid, not necessarily unusual for his time, but his stature lent much weight.

CHAPTER 7

1 Reference to Shakespeare's *As You Like It*, and Jacques, a 'discontented, melancholy lord', who later delivers the famous 'All the world's a stage' monologue.

2 See Pearson (1996).

3 See Haggerty *et al.* (1995).

4 See Pearson and Nicholas (2007).

5 See Pearson (2003).

CHAPTER 8

1 Darwin's grave is placed in the shadow of memorials to William Herschel (1738–1822) and Howard Walter Florey (1898–1968), along the north aisle of the nave leading to the north transept, and alongside the grave of Herschel's son, Sir John Herschel (1792–1871). Memorials to James Clerk Maxwell (1831–1879) and Michael Faraday (1791–1867) separate this group from Sir Isaac Newton's (1642–1727) grave in the central nave, to the front of the choir screen. This area is often referred to as Scientists' Corner.

CHAPTER 9

1 See Gould (1977a) and also Gould (1980).

2 In the popular press, Richard Dawkins devotes a chapter to the issue in *The Blind Watchmaker.* In the scientific literature, Bryan Clarke and Jeffrey Levinton separately denounced punctuated equilibria in the same edition of *Nature* (Clarke 1994 and Levinton 1994), the former recalling a 1949 conference on genetics, palaeontology and evolution when periods with differing evolutionary rates were described as 'storms' and 'doldrums', producing 'unquestionable spurts and bursts, and contrasted long periods of relative stasis'.

3 See Gould (1977b).

4 See Funk *et al.* (2006) and Nosil and Sandoval (2008).

5 In 1909 Danish biologist Wilhelm Johannsen proposed adopting 'the last syllable "gene", which alone is of interest to us, from Darwin's well known word (Pangenesis) and thereby replace the less desirable ambiguous word "determiner." Consequently, we will speak of "the gene" and "the genes" instead of "pangen" and "the pangens." The word gene is completely free from any hypothesis; it expresses only the evident fact that, in any case, many characteristics of the organism are specified in the germ cells by means of special conditions, foundations, and determiners which are present in unique, separate, and thereby independent ways – in short, precisely what we wish to call genes... The "gene" is nothing but a very applicable little word, easily combined with others, and hence it may be useful as an expression for the "unit-factors," "elements" or "allelomorphs" in the gametes, demonstrated by modern Mendelian researches. A "genotype" is the sum total of all the "genes" in a gamete or in a zygote'.

6 Along with his co-authors, particularly Chandra Wickramasinghe, Hoyle proposed that life evolved in space, spreading through the universe via panspermia, arrived on earth as viruses via comets, and continues to do so as the motivation for evolution.

7 Now actually named 'Hoyle's fallacy', misunderstanding the *modus operandi* of evolution as defined by random mutation *with* natural selection led Hoyle to compare the likelihood of complexity (functional proteins) evolving to 'a junkyard contains all the bits and pieces of a Boeing-747, dismembered and in disarray. A whirlwind happens to blow through the yard. What is the chance that after its passage a fully assembled 747, ready to fly, will be found standing there?' During his 'Evolution from Space' Omni Lecture at the Royal Institution (London) on 12 January 1982, Hoyle also coined the term 'intelligent design'.

8 You recalled it very well Brian! From Chapter I, 'Variation under Domestication': 'Professor Wyman has recently communicated to me a good illustration of this fact; on asking some farmers in Virginia how it was that all their pigs were black, they informed him that the pigs ate the paint-root (*Lachnanthes*), which coloured their bones pink, and which caused the hoofs of all but the black varieties to drop off; and one of the "crackers" (*i.e.* Virginia squatters) added, "we select the black members of a litter for raising, as they alone have a good chance of living"'.

CHAPTER 10

1 This is covered in more detail in Gregory (2005), after the original exposition by Lederberg and McCray (2001).

2 Aristotelian essentialism considers biological entities to own an essence (Greek: *ousia*) that encapsulates the qualities that are essential for functioning as that form (see also Chapter 2, n. 1). The duality of dualism arises from the transformation of formless, raw matter into recognisable forms by this essence. This was thought to be the prevalent belief prior to the *Origin of Species*, but this may be a mistruth largely promoted by Ernst Mayr. Its validity is currently the topic of energetic debate; for example, see Mallet (2008).

3 Viewable at http://darwin-online.org.uk/. See also Chapter 4, n. 1.

4 See Mayden (1997) and Wilkins (2002).

5 See Derry (2009d) for the opinions of some leading researchers in this field.

6 See Chapter 3, n. 4.

7 Popular interest in *Encephalitis lethargica*, also known as 'sleepy sickness' or 'sleeping sickness', was reawakened following the 1990 adaptation of Oliver Sacks' *Awakenings* as a film with Robin Williams and Robert De Niro, about varyingly successful treatment of catatonic patients 40 years after the 1920s epidemic responsible for their infection.

8 Darwin might have suffered from Chagas' disease as a result of a bite from the Great Black Bug of the Pampas, reported in his diaries for March 1835.

CHAPTER 11

1 Well, along with Estonia, France, Germany, Lithuania, Poland, the United States, Basque Country, Wales and Serbia.

2 Paget was clearly a troubled individual!

3 A strip of ancient semi-natural woodland. Darwin first leased then bought the land for the Sand-walk from his neighbour John Lubbock.

4 In Chapter III, 'Complex Relations of All Animals and Plants to Each Other in the Struggle for Existence'.

5 A commonly used method of 'fenceline' comparisons for studying the effects of grazing pressure on rangeland quality by excluding an equivalent piece of land from animal impacts.

CHAPTER 12

1 To give credit where it is due, we ought to mention Robert Brout and François Englert who simultaneously theorised the particle's existence. Leon Lederman's book, which coined 'the God particle', was supposed to be titled '*The Goddamn Particle*', but Lederman was overruled.

2 This recapitulates a prediction first made in the television documentary *What We Still Don't Know* (Channel 4, UK, 2004).

CHAPTER 13

1 See Farrell *et al.* (1976).

2 *An Evening with Callow and Fry* (Anglia TV, UK, 2003).

3 From *Creation: Science Confirms the Bible Is True*. Answers in Genesis DVD, 2006.

4 The Confession of Faith includes in it, Chapter IV, 'Of Creation', which states: 'It pleased God the Father, Son, and Holy Ghost, for the manifestation of the glory of his eternal power, wisdom, and goodness, in the beginning, to create, or make of nothing, the world, and all things therein, whether visible or invisible, in the space of six days, and all very good. After God had made all other creatures, he created man, male and female, with reasonable and immortal souls, endued with knowledge, righteousness, and true holiness, after his own image, having the law of God written in their hearts, and power to fulfil it; and yet under a possibility of transgressing, being left to the liberty of their own will, which was subject unto change. Beside this law written in their hearts, they received a command not to eat of the tree of the knowledge of good and evil; which while they kept, they were happy in their communion with God, and had dominion over the creatures'.

5 'Flatereres been the develes chapelleyns, that syngen evere Placebo', from *The Parson's Tale*, the final of Geoffrey Chaucer's *Canterbury Tales*.

6 There is a specific response to this accusation on the *Talk Origins* website under Claim CB102.1: http://www.talkorigins.org/indexcc/CB/CB102_1.html

7 Carl E. Baugh is an American Young Earth creationist and founder of the Creation Evidence Museum, claiming discoveries of coinciding human and dinosaur footprints near the Paluxy River in Texas.

8 Named after the 1857 book *Omphalos*, by Philip Henry Gosse, which argues that God must have created a fully functional universe, with only an appearance of antiquity.

9 See Shiga (2007).

10 The paper is Ronshaugen *et al.* (2002). The Answers in Genesis response is that 'this is really just a straw-man argument because creationists do not claim that mutations cannot alter the body plan. Rather, they claim that mutations do not lead to an increase in information. Indeed, reducing the number of legs may alter the body plan, but it does not explain the origin of legs in the first place'. http://www.answersingenesis.org/articles/2009/01/13/eyes-have-it

11 Ted Haggard famously evicted Richard Dawkins from his ministry car-park, vociferating, 'You called my children animals!' after they had discussed evolution.

CHAPTER 14

1 Now on display at London's Natural History Museum.

2 Much of Ota Benga's story is echoed in the film *Man to Man*, written by William Boyd, he of *Brazzaville Beach*.

3 Wilde is not alone in these sentiments: Cicero (106–43 BC) also wrote, 'Art is born of the observation and investigation of nature', and Marc Chagall (1887–1985), 'Great art picks up where nature ends'.

CHAPTER 15

1 For example, 'Through [human] powers of intellect, articulate language has been evolved; and on this his wonderful advancement has mainly depended' (Darwin 1871).

2 See Holloway *et al.* (2004).

3 See Holloway (2008).

4 See Dunbar and Shultz (2007).

5 *Sensu* Gould and Lewontin (1979).

6 See Atkinson *et al.* (2008) for a recent study.

7 See Chomsky (2005).

8 The second edition was published in 1882.

9 See Christiansen and Kirby (2003).

CHAPTER 16

1 CNN live debate with Michael Shermer on 16 August 2006.

CHAPTER 17

1 Source: *2001 Census* [Key Statistics for Local Authorities]. Crown copyright 2004. Crown copyright material is reproduced with the permission of the Controller of HMSO.

2 Indicated by some overlap in the numbers. An Ipsos MORI poll for BBC's *Horizon* programme in 2006, asking 2000 people what best described their view of the origin and development of life, recorded the following: 22% Creationism, 17% Intelligent Design, 48% evolution theory, 13% didn't know.

3 This is further borne out by a 1997 poll in the USA which found the national split as 44% Creationism, 39% theistic (God-guided) evolution, 10% naturalistic evolution. Among these respondents, the scientific community contributed 5%, 40% and 55%, respectively.

4 TED lecture 'Reverse Engineering and Intelligent Design', February 2006.

5 See Dawkins (1997).

6 From *The Twelve Fundamental Laws of Human Nature*, 13 April 1829.

7 From *Introduction to a Contribution to the Critique of Hegel's Philosophy of Right* written in 1843, but unpublished in his lifetime.

8 Dawkins tells the story that he had wanted to call it *The Root of Nearly*, or *Practically, all Evil*, but his producers favoured the more sensational version.

APPENDIX

1 John's submitted his piece with the title, *Some Specious Readings of The Origin of Species*.

2 Darwin actually introduced Spencer's phrase into the fifth edition of the *Origin of Species* in Chapter III, 'Struggle for Existence', p. 72: '...variations, however slight, and from whatever cause proceeding, if they be in any degree profitable to the individuals of a species, in their infinitely complex relations to other organic beings and to their physical conditions of life, will tend to the preservation of such individuals, and will generally be inherited by the offspring. The offspring, also, will thus have a better chance of surviving, for, of the many individuals of any species which are periodically born, but a small number can survive. I have called this principle, by which each slight variation, if useful, is preserved, by the term Natural Selection, in order to mark its relation to man's power of selection. But the expression often used by Mr. Herbert Spencer of the Survival of the Fittest is more accurate, and is sometimes equally convenient'. And he further sanctions its use with its appearance in the next chapter heading, 'Natural Selection; or the Survival of the Fittest'.

3 See Appleman (1979).

4 See Waddington (1958).

5 See Huxley (1870).

6 See Fisher (1936). As Popper puts it, 'The most careful observation of one developing caterpillar will not help us to predict its transformation into a butterfly'.

7 Julian Huxley, 'Evolutionary Ethics' (Romanes Lecture delivered by Julian Huxley in 1943), in Huxley and Huxley (1947), p. 105.

8 See Peirce (1965). There is a footnote on 'fate': 'Fate means merely that which is sure to come true, and can nohow be avoided. It is superstition to suppose that a certain sort of events are ever fated, and it is another to suppose that the word fate can never be freed from its superstitious taint, "We are all fated to die" '.

9 See Llewelyn (2008).

Glossary of Scientific Terms

Definitions from: the *Encyclopedia of Life Sciences* (Wiley 2006); the *Encyclopedia of Genetics* (Brenner and Miller 2001); the *Oxford English Dictionary*, online at http://www.askoxford.com/; Biology Online at http://www.biology-online.org/; and otherwise as indicated.

Abiogenesis Study of how life originally arose on the planet; encompasses the ancient belief in the spontaneous generation of life from non-living matter.

Adaptation Changes in the function or behaviour of an organism made in response to changes in the external environment.

Algorithm Set of rules for solving a problem in a finite number of steps.

Allele One of two or more alternative forms of a gene that can occupy the same position, or gene locus, on the chromosome.

Allopatric Occupying separate geographic areas.

Altruism Behaviour of an animal that benefits another at its own expense.

Artificial selection Commonly known as selective breeding, where professionals study the genotype and phenotype of parent organisms in the hope of producing a hybrid that possesses many of the desirable characteristics found in its parents.

Assortative mating Tendency for the characteristics of mating partners to be correlated. For example, large males may tend to mate with large females (positive assortative mating) or pale-coloured males with dark-coloured females (negative assortative mating).

Backcross Mating between a hybrid organism and one of its parents.

Chromosome Thread-like structure found in the nuclei of most living cells, carrying genetic information in the form of genes.

Coalescent theory Retrospective model of population genetics. It employs a sample of individuals from a population to trace all alleles of a gene shared by all members of the population to a single ancestral copy, known as the most recent common ancestor (MRCA; sometimes also termed the coancestor to emphasise the coalescent relationship) (Nordborg 2001).

Cross-mating To interbreed.

Crossover See recombination.

Diffusion theory Mathematical description of the process of diffusion, used to describe how deterministic forces, such as mutation, selection and migration, and the stochastic process of genetic drift interact to influence the fate of mutations in populations (see Watterson 1996).

Diploid Individual or cell having two full sets of chromosomes. The number of chromosomes is species specific.

Divergence Process of one species diverging over time into more than one species.

DNA Deoxyribonucleic acid. The biological macromolecule that carries the permanent store of genetic information in a cell. It is a polymer of the nucleotides adenine, cytosine, guanine and thymine, and the genetic information is carried in the form of the sequence of nucleotides, which corresponds to a sequence of amino acids in protein. DNA is transcribed to make messenger RNA, which is used to direct the synthesis of proteins.

DNA sequencing Methods for determining the order of the nucleotide bases (adenine, guanine, cytosine and thymine) in a molecule of DNA.

Epigenetic Resulting from external rather than genetic influences.

Eukaryotes Organisms of the Domain Eukarya: protists, animals and plants. They may be multicellular or unicellular and have cells with a complex organisation, distinguished from the cells of prokaryotes by a distinct nucleus containing the DNA in the form of chromosomes, and a cytoplasm containing membrane-bounded organelles of different functions such as mitochondria, endoplasmic reticulum, Golgi bodies and (in plants) chloroplasts.

Evolutionary computation Subfield of artificial intelligence that uses techniques inspired by evolutionary biology such as inheritance, mutation, selection and recombination.

Evolutionary developmental biology Comparison of the developmental processes of different organisms in order to identify the ancestral relationship between them, as well as how developmental processes evolved.

Filial Relating to or due from a son or daughter.

Gamete Haploid germ cell produced by a sexually reproducing organism. A gamete is capable of fusing with another gamete of opposite sex to produce a zygote. The male gamete is usually called the sperm, and the female the ovum.

Gene Classically, the region of a chromosome that controls a single hereditary trait. The term *gene* is sometimes used as a synonym for allele (a particular version of a gene) and sometimes as a synonym for locus (the position on the chromosome occupied by the gene). At the molecular level, a gene is a sequence of DNA that

155

encodes the information for a protein or an RNA, together with the regulatory sequences necessary for the gene's expression.

Gene exchange Transfer of alleles between populations or species by hybridisation followed by backcrossing to the parental forms. This process is equivalent to gene flow but the term *gene exchange* is usually used to refer to interchange between genetically distinct populations.

Gene expression Transcription and/or translation of a gene to form the RNA and/or protein product. Regulation of gene expression can be at the level of transcription or translation.

Gene flow Exchange of genes between populations caused by immigration and emigration of individuals across populations.

Gene pool (1) All the alleles present in a population of sexually reproducing individuals. (2) All alleles at a given locus in such a population.

Genetic drift Random changes in allele frequencies in small isolated populations as a result of factors other than natural selection, such as sampling of only small numbers of gametes in each generation.

Genome Total set of genes, and any additional non-coding DNA, carried by an individual organism, cell or virus.

Genotype Genetic make-up of an organism.

Germ cell Progenitor of a gamete.

Germ plasm Component of germ cells that Weismann proposed were responsible for heredity, roughly equatable to our modern understanding of DNA.

Gradualism Gradual rather than sudden change.

Group selection Differential survival or reproduction of groups of organisms, such as a population of parasites on a host.

Haploid Individual or cell (e.g. a gamete) having a single set of chromosomes. The number of chromosomes is species specific.

Homeobox Highly conserved DNA sequence, contained in a gene (e.g. HOX gene), coding for a protein that is highly conserved, and can bind to DNA to control gene expression.

Homologous (1) DNA sequences, molecules or structures that are similar as a result of their derivation from a common ancestor. (2) Maternal and paternal copies of a given chromosome in a diploid organism, which carry the same genetic loci although they may carry different alleles.

Horizontal gene transfer (1) Transmission of genetic material between cells in a process not involving reproduction. (2) Transfer of genetic material between species.

HOX genes Homeobox-containing genes that are characteristic of animals and are involved in specifying positional identity along the anterior–posterior axis.

Hybrid (1) Offspring of a cross between two pure-breeding lines of different genotype. (2) Offspring of a cross between different species.

Hybrid sterility Sterility of hybrids; for example, a mule (a hybrid between a female horse and male donkey), which can develop into an adult but fails to develop functional gametes, and is therefore sterile.

Hybrid zone Region where the geographic ranges of two genetically divergent groups of populations meet and produce at least some offspring of mixed ancestry.

Hybridisation Cross between individuals from genetically differentiated populations.

Identity by descent Two genes are identical by descent if, and only if, they are derived from the same gene, or one is derived from the other (in both cases without mutation) (Nagylaki 1989).

Inheritable trait Character capable of being inherited, for example human eye colour.

Instinct Inherent disposition of a living organism towards a particular behaviour, where sequences of behaviour are unlearned and inherited.

Lamarckian inheritance Once widely accepted idea that an organism can pass on characteristics that it acquired during its lifetime to its offspring.

Locus Position on a chromosome at which the gene controlling a particular trait resides. It can be occupied by different alleles of the gene.

Macroevolution Large-scale evolution, usually interpreted to mean all processes and patterns of phenotypic and genetic change at and above the species level.

Macromutation Simultaneous mutations occurring at the same time, or when a sudden large-scale mutation produces a characteristic.

Microevolution Small-scale changes in allele frequencies in a population, thus change at or below the species level.

Mutation Permanent heritable change in the genetic material of a cell or organism. This could be a change in the base sequence of the DNA that affects a single gene, or a change in the number or structure of the chromosomes. In classical genetics, the term usually refers to a change that has a demonstrable effect on the phenotype.

Natural selection Evolutionary process whereby organisms better adapted to their environment tend to survive and produce more offspring.

Neo-Lamarckism Modern version of *Lamarckian inheritance*, emphasising the importance of environmental factors in genetic changes and retaining the notion of the inheritance of acquired characters.

Nucleotides Units of DNA and RNA: mono-, di- and triphosphates of adenosine, guanosine, thymidine, uridine, cytidine, deoxyadenosine, deoxyguanosine, deoxythymidine, deoxyuridine and deoxycytidine.

Pangenesis Darwin's hypothetical mechanism for heredity where gemmules are shed by the body cells and carried in the bloodstream to the gonads, where they accumulate within the reproductive cell (e.g. egg) prior to fertilisation.

Parapatric Occupying geographic areas that are contiguous but not overlapping.

Phenotype Morphological, biochemical and behavioural characteristics of an individual organism or cell, determined by both genetic and environmental factors.

Population genetics Adaptation and speciation studies of the allele frequency and change under the influence of natural selection, genetic drift, mutation and gene flow, taking account of population distribution.

Punctuated equilibrium. In contrast to gradualism, relates most evolutionary change to rapid divergence, with periods of relative stasis between speciations.

Quantitative trait locus (QTL) Not necessarily a gene, but a region of DNA associated with continuous phenotypic traits (those traits that vary continuously, e.g. height) as opposed to discrete traits (traits that have two or several character values, e.g. eye colour in humans or smooth versus wrinkled peas used by Mendel in his experiments). The QTL number can reveal, for example, that plant height is controlled by many genes of small effect, or by a few genes of large effect.

Random genetic drift See *genetic drift*.

Recombination (1) Exchange of similar or dissimilar sequences between two DNA molecules. (2) Any process that gives rise to cells or individuals in which parental alleles have been inherited in new combinations.

Reinforcement Process by which natural selection increases reproductive isolation.

Reproductive isolation Separation of gene pools preventing two or more populations from exchanging genes, either by preventing fertilisation, or via hybrid sterility.

Retroposon Repetitive DNA fragments inserted into chromosomes after being reverse transcribed from any RNA molecule.

Reverse transcription Synthesis of DNA using RNA as the template.

RNA Ribonucleic acid. Nucleic acid, usually single-stranded except in some virus genomes, which is composed of ribonucleotides. The bases in RNA are adenine, cytosine, guanine and uracil.

Saltation Sudden rather than gradual change.

Sexual selection Form of natural selection where the male or female is attracted by certain characteristics (form, colour, behaviour, etc.) in the opposite sex.

Soma (1) Animal body, with the exception of cells of the germline. (2) Neuronal cell body containing cell nucleus.

Somatic hypermutation Accumulation of point mutations that occurs in the variable regions of antibody genes.

Species selection Macroevolutionary processes which shape evolution at and above the level of species and are not driven by the microevolutionary mechanisms that are the basis of the modern synthesis.

Sympatric Occupying the same geographic region, thus having the potential to interbreed.

Taxon (pl. taxa) A group such as Mammalia or *Homo sapiens* that is a unit, rather than a kind of unit (e.g. family, class), in a formal classification system.

Transcription The synthesis of RNA molecules using DNA as the template to determine the sequence of bases in the RNA product.

Transcription factor A protein required to initiate, regulate or repress transcription.

Translation Process whereby the nucleotide sequence of a messenger RNA is 'read out' and used to make a polypeptide.

Transposable element DNA sequence able to move itself, or a copy of itself, within the genome, often causing mutations.

Zoophyte Plant-like animal, especially a coral, sea anemone, sponge, or sea lily.

Bibliography

A note on sourcing Darwin's works

The excellent and ongoing Darwin Correspondence Project has reached 1869 (volume 17, Cambridge University Press; http://www.darwinproject.ac.uk/) at the time of writing, but where should one look to obtain the complete works that document Darwin's life research?

Electronic Several vendors offer Darwin's complete works as e-books on CD-ROM (for example, these can be bought on eBay). Prices vary from 50p to £5. Digital texts of the complete works are available for free at *The Complete Work of Charles Darwin Online* (http://darwin-online.org.uk/).

Bound Pickering & Chatto probably offer the best quality when purchasing a set of Darwin's complete works in 29 matching cloth-bound and admirably presented volumes, available as a whole set or separately, although this is possibly the only source of the complete works as a single matching set. Details are available at http://www.pickeringchatto.com/.

Selected The Collector's Library of Essential Thinkers publish *Darwin: Selected Works*, and Orion have released the audio book *A Darwin Selection: Understanding Darwin*, edited by Mark Ridley and narrated by Sir Derek Jacobi. Both are available through Amazon.

Various Penguin re-release individual works intermittently, recently *The Expression of Emotions...* and *Autobiographies*. Icon Books has also recently published an attractive version of *The Autobiography of Charles Darwin*. Since 2004 Kessinger Publishing Co. have offered most of Darwin's works either bound or electronically. Indypublish.com offer most of Darwin's works exiguously printed on demand in minimalist paperback and hardback, but these are reported to lack original diagrams. Lightning Source UK Ltd have recently released *Geological Observations...* and Fontana Press have published their own attractive version of *The Expression of the Emotions....* The

University Press of the Pacific have competently published facsimiles of *The Various Contrivances...* and *The Movements and Habit....* Numerous publishers (e.g. Vintage Classics, W.W. Norton & Co. Ltd, Cambridge University Press, Oxford University Press, Sterling, Arcturus Publishing Ltd, Pocket Books, Penguin Books Ltd, Barnes & Noble, Grove Press, Wilder Publications Ltd, Castle Books, Gramercy Books, Signet Classics, Everyman's Library, Harvard University Press, Oxford World's Classics) have or soon will have released their editions of the versions of *On the Origin of Species* or *Voyage of the Beagle* with varying introductions and notes. BiblioBazaar and ReadHowYouWant.com offer large-print editions.

Darwin's major published books and chapters

Reproduced with permission from John van Wyhe (ed.) *The Complete Work of Charles Darwin Online* (http://darwin-online.org.uk/).

Darwin, C.R. 1829–32. [*Records of captured insects*]. In Stephens, *Illustrations of British entomology.*

_____ 1835. *Extracts from letters to Professor Henslow* (private printing).

_____ 1838–43. *The zoology of the voyage of H.M.S. Beagle.* Edited and superintended by Charles Darwin. (Original issue in 5 parts of 19 numbers.)
  • *Fossil Mammalia* by R. Owen. Includes by Darwin: Preface & Geological introduction.
  • *Mammalia* by G.R. Waterhouse. Includes by Darwin: Geographical introduction & A notice of their habits and ranges.
  • *Birds* by J. Gould [and G.R. Gray].
  • *Fish* by L. Jenyns.
  • *Reptiles [and Amphibia]* by T. Bell.

_____ 1839. *The narrative of the voyages of H.M. Ships Adventure and Beagle* (1st ed.), 3 vols and appendix:
  • *Proceedings of the first expedition*, 1826–30 by P.P. King.
  • *Proceedings of the second expedition*, 1831–36 by R. FitzRoy.
  • *Appendix* by R. Fitzroy.
  • *Journal and remarks*, 1832–36 by C.R. Darwin.

_____ 1842. *The Geology of the Voyage of the Beagle, Under the Command of Capt. Fitzroy, R.N. During the Years 1832 to 1836. First Part: The Structure and Distribution of Coral Reefs.*

_____1844. *The Geology of the Voyage of the Beagle, Under the Command of Capt. Fitzroy, R.N. During the Years 1832 to 1836. Second Part: Geological observations on the volcanic islands visited during the voyage of H.M.S. Beagle, together with some brief notices of the geology of Australia and the Cape of Good Hope.*

_____ 1844. [Extracts from letters on guanacos]. In Walton, W., *The alpaca.*

_____ 1845. [Testimonial]. In Brayley, E.W., *Additional testimonials submitted to the Council of University College, London.*

_____ 1846. [Query on coral reefs]. In Stokes, J.L., *Discoveries in Australia.*

_____ 1846. *The Geology of the Voyage of the Beagle, Under the Command of Capt. Fitzroy, R.N. During the Years 1832 to 1836. Third Part: Geological Observations on South America.*

_____ 1849. *Geology from A Manual of scientific enquiry; prepared for the use of Her Majesty's Navy: and adapted for travellers in general.* (John F.W. Herschel, ed.)

_____ 1851. [Testimonial]. In *Testimonials for Thomas H. Huxley, F.R.S., candidate for the Chair of Natural History at the University of Toronto.*

_____ 1851. *A Monograph of the Sub-class Cirripedia, with Figures of all the Species. Vol. 1 The Lepadidae; or, Pedunculated Cirripedes.*

_____ 1851. *A monograph on the fossil Lepadidae, or, pedunculated cirripedes of Great Britain.*

_____ 1852. [Letter on the bookselling question]. In Parker, J., *The opinions of certain authors on the bookselling question.*

_____ 1854. *A Monograph of the Sub-class Cirripedia, with Figures of all the Species. Vol. 2 The Balanidae (or Sessile Cirripedes); the Verrucidae, etc.*

_____ 1854. *A monograph on the fossil Balanidae and Verrucidae of Great Britain.*

_____ 1858. *On the Tendency of Species to form Varieties; and on the Perpetuation of Varieties and Species by Natural Means of Selection* (Extract from an unpublished Work on Species).

_____ 1859. *On the Origin of Species by Means of Natural Selection, or the Preservation of Favoured Races in the Struggle for Life.*

_____ 1862. [Query to Army Surgeons.]

_____ 1862. *On the Various Contrivances by which British and Foreign Orchids are Fertilised by Insects.*

_____ 1862. [Recollections of Professor Henslow]. In Jenyns, L., *Memoir of the Rev. John Stevens Henslow.*

Darwin, Emma and _____ 1863. *An appeal [against steel vermin traps].*

_____ 1865. [Testimonial]. In *Testimonials in favour of Mr. Adam White.*

_____ 1867. Queries about expression. In Freeman, R.B. and Gautrey, P.J. (eds.), *Charles Darwin's Queries about expression.*

_____ 1868. *The Variation of Plants and Animals Under Domestication.*

_____ 1870. [Note on the age of certain birds]. In Lankester, E.R., *On comparative longevity in man and the lower animals*, p. 58.

_____ 1871. *The Descent of Man, and Selection in Relation to Sex.*

_____ 1872. *The Expression of the Emotions in Man and Animals.*

_____ 1873. [Testimonial]. In *Testimonials in favour of W. Boyd Dawkins.*

_____ 1874. Physiognomy. In *Notes and queries on Anthropology, for the use of travellers and residents in uncivilized lands.*

_____ 1875. [Letter to Haeckel on the origins of Darwin's theory of evolution]. In Schmidt, O., *The doctrine of descent and Darwinism.*

_____ 1875. *The movements and habits of climbing plants.*

_____ 1875. *Insectivorous Plants.*

_____ 1876. *The Effects of Cross and Self Fertilisation in the Vegetable Kingdom.*

_____ 1876. [Evidence given to the Commission]. *Report of the Royal Commission on the practice of subjecting live animals to experiments for scientific purposes.*

_____ 1877. *The Different Forms of Flowers on Plants of the Same Species.*

_____ 1877. *To members of the Down Friendly Club.*

_____ 1878. Prefatory letter. In Kerner, A., *Flowers and their unbidden guests.*

_____ 1879. "Preface and 'a preliminary notice'" in Ernst Krause's *Erasmus Darwin.*

_____ 1880. *The Power of Movement in Plants.*

_____ 1881. *The Formation of Vegetable Mould, Through the Action of Worms.*

_____ 1881. [Extracts from 2 letters on the drift deposits near Southampton]. In Geikie, J., *Prehistoric Europe.*

_____ 1881. *Correspondence with Charles Darwin LL.D., F.R.S., on experimenting upon living animals.*

_____ 1882. Prefatory notice. In Weismann, A., *Studies in the theory of descent.*

_____ 1882. In Romanes, G.J., *Animal intelligence.*

_____ 1883. Prefatory notice. In Müller, H., *The fertilisation of flowers.*

_____ 1883. Essay on instinct. In Romanes, G.J., *Mental evolution in animals.*

_____ 1887. *The life and letters of Charles Darwin, including an autobiographical chapter.* (Francis Darwin, ed.)

_____ 1909. *The foundations of The origin of species, a sketch written in 1842.* (Francis Darwin, ed.)

_____ 1909. *The foundations of The origin of species. Two essays written in 1842 and 1844.* (Francis Darwin, ed.)

_____ 1958. *Autobiography of Charles Darwin.* (Barlow, unexpurgated)

General reference and further reading

Here you will recognise the names of some of the contributors to this book. I wholeheartedly recommend getting hold of a copy of their books and articles in order to better understand their points of view. There is a good deal of science in this book; the subject matter necessitates as much, if not more. Having extolled the virtues of the scientific method I am bound to refer to at least some of the work that I have drawn upon. This list also includes recommendations for further reading.

Allen, G. 1897. *The Evolution of the Idea of God.* The Book Tree, San Diego, CA, 2000.

Appleman, P. (ed.) 1979. Charles Darwin, *The Descent of Man*, second edition, part I, chapter II. In *Darwin: A Norton Critical Edition.* W.W. Norton & Co., New York, p. 173.

Ashworth, J.H. 1935. Charles Darwin as a student in Edinburgh, 1825–1827. *Proceedings of the Royal Society of Edinburgh* 55: 97–113.

Atkinson, Q.D., Meade, A., Venditti, C., Greenhill, S.J. and Pagel, M. 2008. Languages evolve in punctuational bursts. *Science* 319: 588.

Avery, J. 2003. *Information Theory and Evolution.* World Scientific, Singapore.

Baggini, J. 2003. *Atheism.* Oxford University Press, Oxford.

Baldwin, J.M. 1896. A new factor in evolution. *American Naturalist* 30: 441–451, 536–553.

Barnett, S.A. (ed.) 1958. *A Century of Darwin.* William Heinemann Ltd, London.

Barton, N.H., Briggs, D.E.G., Eisen, J.A., Goldstein, D.B. and Patel, N.H. 2007. *Evolution.* Cold Spring Harbor Laboratory Press, New York.

Benitez-Bribiesca, L. 2001. Memetics: A dangerous idea. *Interciencia* 26: 29–31.

Berry, A. (ed.) 2002. *Infinite Tropics: An Alfred Russel Wallace Anthology.* Verso, London.

Bhaktivedanta, A.C. (trans.) 1968. *Bhagavad-Gītā As It Is.* Macmillan Publishers, London.

Blackmore, S. 1999. *The Meme Machine.* Oxford University Press, Oxford.

Blackmore, S. 2005. *Consciousness.* Oxford University Press, Oxford.

Boyd, W. 1990. *Brazzaville Beach.* Sinclair-Stevenson Ltd, London.

Brenner, S. and Miller, J. 2001. *Encyclopedia of Genetics.* Elsevier, Amsterdam.

Browne, J. 2003a. *Charles Darwin. Voyaging. Volume I of a Biography.* Pimlico, London.

Browne, J. 2003b. *Charles Darwin. The Power of Place. Volume II of a Biography.* Pimlico, London.

Bulmer, M. 2004. Did Jenkin's swamping argument invalidate Darwin's theory of natural selection? *British Journal for the History of Science* 37: 281–297.

Buss, D.M. (ed.) 2005. *Handbook of Evolutionary Psychology.* John Wiley & Sons, Chichester.

Butt, K.R. and Lowe, C.N. 2005. *Earthworms at Down House.* Report to English Heritage, University of Central Lancashire, Preston, Lancashire.

Calvin, W.H. and Bickerton, D. 2000. *Lingua ex Machina: Reconciling Darwin and Chomsky with the Human Brain.* MIT Press, Cambridge, MA.

Cangelosi, A. and Parisi, D. 2002. *Simulating the Evolution of Language.* Springer-Verlag, London.

Carroll, S.B. 2006. *Endless Forms Most Beautiful.* Weidenfeld & Nicolson, London.

Chambers, R. 1844. *Vestiges of the Natural History of Creation and Other Evolutionary Writings.* John Churchill, London. University of Chicago Press, Chicago, IL, 1994.

Chapman, M. 2000. *Trials of the Monkey: An Accidental Memoir.* Duckworth, London.

Charlesworth, B. 1982. Neo-Darwinism – the plain truth. *New Scientist* 94: 133–137.

Charlesworth, B. and Charlesworth, D. 2003. *Evolution.* Oxford University Press, Oxford.

Charnov, E.L. 1976. Optimal foraging: The marginal value theorem. *Theoretical Population Biology* 9: 129–136.

Chomsky, N. 1965. *Aspects of the Theory of Syntax.* MIT Press, Cambridge, MA.

Chomsky, N. 2005. Some simple evo-devo theses: How true might they be for language? *Alice V. and David H. Morris Symposium on Language and Communication;*

The Evolution of Language, 14 October 2005, Stony Brook University, New York. Available online at: http://www.punksinscience.org/kleanthes/courses/MATERIALS/Chomsky_Evo-Devo.doc

Christiansen, M.H. and Kirby, S. 2003a. Language evolution: consensus and controversies. *Trends in Cognitive Sciences* 7: 300–307.

Christiansen, M.H. and Kirby, S. (eds.) 2003b. *Language Evolution*. Oxford University Press, Oxford.

Clarke, B. 1994. Punctuated debate. *Nature* 368: 407.

Conway Morris, S. 1998. *The Crucible of Creation*. Oxford University Press, Oxford.

Conway Morris, S. 2003. *Life's Solution*. Cambridge University Press, Cambridge, UK.

Cork, B. and Bresler, L. 1985. *Evolution*. Usborne, London.

Cornwall, J. 2007. *Darwin's Angel: An Angelic Repost to The God Delusion*. Profile Books Ltd, London.

Coyne, J.A. and Orr, H.A. 2004. *Speciation*. Sinauer Associates, Sunderland, MA.

Dalrymple, W. 1998. *The Age of Kali*. Harper Collins, London.

Darwin, F. *c.*1884. [Preliminary draft of] *Reminiscences of My Father's Everyday Life*. In Darwin, F. 1929. *The Autobiography of Charles Darwin* [Icon Books Ltd, Cambridge, UK, 2003].

Davis, A.S. 1871. The "North British Review" and the Origin of Species. *Nature* 5: 161.

Davis, P. and Kenyon, D.H. 1989. *Of Pandas and People: The Central Question of Biological Origins*. Foundation for Thought and Ethics, Richardson, TX.

Dawkins, R. 1976. *The Selfish Gene*. Oxford University Press, Oxford.

Dawkins, R. 1982. *The Extended Phenotype*. Oxford University Press, Oxford.

Dawkins, R. 1997. Is Science a Religion? *The Humanist* 57: 1.

Dawkins, R. 2003. *A Devil's Chaplain*. Weidenfeld & Nicolson, London.

Dawkins, R. 2004. *The Ancestor's Tale: A Pilgrimage to the Dawn of Life*. Weidenfeld & Nicolson, London.

Dawkins, R. 2006. *The God Delusion*. Bantam Press, London.

Dawood, N.J. (trans.) 1956. *The Koran*. Penguin Books Ltd, London.

Dembski, W.A. 2004. *The Design Revolution: Answering the Toughest Questions about Intelligent Design*. InterVarsity Press, Downers Grove, IL.

Dennett, D.C. 1991. *Consciousness Explained*. Little, Brown & Co., London.

Dennett, D.C. 1995. *Darwin's Dangerous Idea: Evolution and the Meanings of Life*. Penguin Books Ltd, London.

Dennett, D.C. 1998. *Brainchildren: Essays on Designing Minds (Representation and Mind)*. MIT Press, Cambridge, MA.

Dennett, D.C. 2006. *Breaking the Spell: Religion as a Natural Phenomenon*. Penguin Books Ltd, London.

Derry, J.F. 2004. (Book Review) What Is Life? By Erwin Schrödinger. *Human Nature Review* 4: 124–125.

Derry, J.F. 2009a. Darwin and the University of Edinburgh. *Edinburgh Science Magazine* 2: 28–29.

Derry, J.F. 2009b. Darwin in disguise. *Trends in Ecology and Evolution* 24: 73–79.

Derry, J.F. 2009c. Bravo Darwin! Emma's pianos and music in the home and work of Charles Darwin. *Endeavour* 33: 35–38.

Derry, J.F. 2009d. Does genomics need Darwin? *Genomics Forum Bulletin* 9 (April).

Derry, J.F. 2009e. Darwin's Edinburgh connection. *Surgeon's News* (July), Edinburgh Royal College of Surgeons.

Derry, J.F. 2009f. Thinking Kong. *Edinburgh Science Magazine* 4:26

Derry, J.F. and Dougill, A.J. 2008. Water availability, piospheres and evolution in African ruminants. *African Journal of Range and Forage Science* 25: 79–92.

Desmond, A. 1984. Robert E. Grant: The social predicament of a pre-Darwinian transmutationist. *Journal of the History of Biology* 17: 189–223.

Desmond, A. 1994. *Huxley: The Devil's Disciple*. Michael Joseph Ltd, London.

Desmond, A. and Moore, J. 1991. *Darwin*. Michael Joseph Ltd, London.

Dunbar, R.I.M. and Shultz, S. 2007. Evolution in the social brain. *Science* 317: 1344–1347.

Dupré, J. 2003. *Darwin's Legacy: What Evolution Means Today*. Oxford University Press, Oxford.

Dyson, G. 1997. *Darwin Among the Machines*. Allen Lane, London.

Eldridge, N. 2005. *Darwin: Discovering the Tree of Life*. W.W. Norton & Co., New York.

Eldridge, N. 2009. What Darwin learned in medical school. *The Lancet* 373: 454–455.

Endersby, J. 2007. Creative designs? How Darwin's Origin caused the Victorian crisis of faith, and other myths. *Times Literary Supplement* (16 March).

Farrell, E.J., Gage, T.E., Pfordresher, J. and Rodrigues, R.J. (eds.) 1976. *Reality in Conflict: Literature of Values in Opposition*. Scott, Foresman and Company, Glenview, IL.

Fisher, H.A.L. 1936. *History of Europe. Vol. I: Ancient and Mediaeval*. Edward Arnold, London, p. vii. Cited in Popper (1961), p. 109.

Fisher, R.A. 1915. The evolution of sexual preference. *Eugenics Review* 7: 184–192.

Fisher, R.A. 1930. *Genetical Theory of Natural Selection*. Clarendon Press, Oxford.

Focke, W.O. 1881. *Die Pflanzen-Mischlinge: Ein Beitrag zur Biologie der Gewächse* (The plant hybrids: a contribution to the biology of plants). Borntraeger, Berlin.

Ford, D.F. 1999. *Theology*. Oxford University Press, Oxford.

Fraser, A.G. 1989. *The Building of Old College: Adam, Playfair and the University of Edinburgh*. Edinburgh University Press, Edinburgh.

Funk, D.J., Nosil, P. and Etges, W.J. 2006. Ecological divergence is consistently positively associated with reproductive isolation across disparate taxa. *Proceedings of the National Academy of Sciences USA* 103: 3209–3213.

Gosse, P.H. 1857. *Omphalos: An Attempt to Untie the Geological Knot*. John Van Voorst, London.

Gould, S.J. 1977a. Evolution's erratic pace. *Natural History* 86: 12–16.

Gould, S.J. 1977b. The return of hopeful monsters. *Natural History* 86: 22–30.

Gould, S.J. 1980. The episodic nature of evolutionary change. In *The Panda's Thumb*. W.W. Norton & Co., New York, pp. 182–184.

Gould, S.J. 1985a. Fleeming Jenkin revisited; this obscure, but able, Victorian gentleman convinced Darwin himself on an important evolutionary point. *Natural History* (June).

Gould, S.J. 1985b. *The Flamingo's Smile*. W.W. Norton & Co., New York.

Gould, S.J. 1990. *Wonderful Life: The Burgess Shale and the Nature of History*. Hutchinson Radius, London.

Gould, S.J. 2000. *The Lying Stones of Marrakech: Penultimate Reflections in Natural History*. Jonathan Cape, London.

Gould, S.J. 2001. *Rocks of Ages: Science and Religion in the Fullness of Life*. William Heinemann, London.

Gould, S.J. 2002a. *I Have Landed: Splashes and Reflections in Natural History*. Jonathan Cape, London.

Gould, S.J. 2002b. *The Structure of Evolutionary Theory*. Harvard University Press, Cambridge, MA.

Gould, S.J. 2003. *The Hedgehog, the Fox, and the Magistrate's Pox*. Jonathan Cape, London.

Gould, S.J. and Lewontin, R.C. 1979. The spandrels of San Marco and the Panglossian paradigm: a critique of the adaptationist programme. *Proceedings of the Royal Society of London B* 205: 581–598.

Grafen, A. and Ridley, M. (eds.) *Richard Dawkins: How a Scientist Changed the Way We Think*. Oxford University Press, Oxford.

Grammer, K., Renninger, L. and Fischer, B. 2004. Disco clothing, female sexual motivation, and relationship status: is she dressed to impress? *Journal of Sex Research* 41: 66–74.

Greene, J.C. 1963. *Darwin and the Modern World View*. Mentor, New York.

Gregory, T.R. 2005. Genome size evolution in animals. In *The Evolution of the Genome*. Elsevier, Amsterdam, pp. 3–87.

Gross, D.C. 2004. Hippocampus minor and man's place in nature: a case study in the social construction of neuroanatomy. *Hippocampus* 3: 403–415.

Haggerty, J.A., Premoli Silva, I., Rack, F. and McNutt, M.K. (eds.) 1995. *Proceedings of the Ocean Drilling Program, Scientific Results* 144: 411–418.

Ham, K. (ed.) 2006. *The New Answers Book: Over 25 Questions on Creation/Evolution and the Bible*. Answers in Genesis, Petersburg, KY.

Hammerstein, P. and Hagen, E.H. 2005. The second wave of evolutionary economics in biology. *Trends in Ecology and Evolution* 20: 604–609.

Henderson, B. 2006. *The Gospel of the Flying Spaghetti Monster*. Harper Collins, London.

Henry, H. 2006. Aristotle on the mechanism of inheritance. *Journal of the History of Biology* 39: 425–455.

Hitchens, C. 2007. *God is Not Great: The Case Against Religion*. Atlantic Books, London.

Henig, R.M. 2000. *A Monk and Two Peas: The Story of Gregor Mendel and the Discovery of Genetics*. Weidenfeld & Nicolson, London.

Hoffmann, H. 1869. *Untersuchungen zur Bestimmung des Werthes von Species und Varietät* (Examinations to determine the value of species and variety). J. Richter, Gießen.

Holland, J. 1975. *Adaptation in Natural and Artificial Systems*. University of Michigan Press, Ann Arbor, MI.

Holloway, R.L. 2008. The human brain evolving: a personal retrospective. *Annual Review of Anthropology* 37. Review in Advance: doi:10.1146/annurev. anthro.37.081407.085211.

Holloway, R.L., Broadfield, D.C., Yuan, M.S., Schwartz, J.H. and Tattersall, I. 2004. *The Human Fossil Record. Vol. 3: Brain Endocasts – The Paleoneurological Evidence*. Wiley-Liss, New York.

Honeywill, R. 2008. *Lamarck's Evolution: Two Centuries of Genius and Jealousy*. Murdoch Books, Sydney.

Hope, J. and van Loon, B. 1994. *Buddha*. Icon Books Ltd, Cambridge, UK.

Houston, R.A. 2002. *Scottish Literacy and the Scottish Identity: Illiteracy and Society in Scotland and Northern England, 1600–1800*. Cambridge Studies in Population, Economy and Society in Past Time (No. 4). Cambridge University Press, Cambridge, UK.

Howard, J. 1982. *Darwin*. Oxford University Press, Oxford.

Hume, D. 1739–40. *A Treatise of Human Nature: Being an Attempt to Introduce the Experimental Method of Reasoning into Moral Subjects*. 3 vols. Vols 1 and 2: John Noon, London, 1739; vol. 3: Thomas Longman, London, 1940.

Hume, D. 1748. *Philosophical Essays Concerning Human Understanding*. A. Millar, London. Revised edition: M. Cooper, London, 1751; republished as *An Enquiry Concerning Human Understanding in Essays and Treatises on Several Subjects*. A. Millar, London/A. Kincaid & A. Donaldson, Edinburgh, 1758.

Hume, D. 1751. *An Enquiry Concerning the Principles of Morals*. A. Millar, London.

Hume, D. 1757. *The Natural History of Religion*. London.

Hume, D. 1779. *Dialogues Concerning Natural Religion*. Robinson, London.

Hutton, J. 1788. *Theory of the Earth; or an Investigation of the Laws observable in the Composition, Dissolution, and Restoration of Land upon the Globe. Transactions of the Royal Society of Edinburgh, Vol. 1, Part 2, pp. 209–304.* Kessinger Publishing Co., Whitefish, MT, 2004.

Huxley, J. 1942. *Evolution: The Modern Synthesis*. Allen & Unwin, London.

Huxley, J., Hardy, A.C. and Ford, E.B. 1954. *Evolution as a Process*. Allen & Unwin, London.

Huxley, T.H. 1863. *Evidence as to Man's Place in Nature*. D. Appleton & Co, New York.

Huxley, T.H. 1870. Geological contemporaneity and persistent types of life. In *Lay Sermons, Addresses, and Reviews*. Macmillan, London, pp. 236–237. Cited in Popper (1961), p. 108.

Huxley, T.H. and Huxley, J. 1947. *Evolution and Ethics 1893–1943*. Pilot Press, London.

Irons, J.C. (ed.) 1896. *Autobiographical sketch of James Croll … with memoir of his life and work*. E. Stanford, London.

Jenkin, F. 1867. (Review of) The Origin of Species. *The North British Review* 46 (June): 277–318.

Jones, S. 1993. *The Language of the Genes*. Harper Collins, London.

Jones, S. 1999. *Almost Like a Whale*. Doubleday, London.

Keegan, R.T. 1996. Getting started: Charles Darwin's early steps toward a creative life in science. *Journal of Adult Development* 3: 7–20.

Keynes, R. 2001. *Annie's Box: Charles Darwin, His Daughter and Human Evolution.* Fourth Estate, London.

Keynes, R.D. 1997. Steps on the path to the Origin of Species. *Journal of Theoretical Biology* 187: 461–471.

Kinsey, A.C., Pomeroy, W.B., Martin, C.E. and Gebhard, P.H. 1953. *Sexual Behavior in the Human Female.* W.B. Saunders, Philadelphia, PA.

Kirby, S. 1999. *Function, Selection and Innateness: The Emergence of Language Universals.* Oxford University Press, Oxford.

Kirkpatrick, M. and Ryan, M.J. 1991. The evolution of mating preferences and the paradox of the lek. *Nature* 350: 33–38.

Kitcher, P. 2007. *Living with Darwin.* Oxford University Press, Oxford.

Kohn, M. 2004. *A Reason for Everything.* Faber & Faber Ltd, London.

Lacroix-Hopson, E. 2001. *Creation, Evolution and Eternity: A Bahá'í's Perspective on Religion and Science.* Yachay Wasi, Inc., New York.

Lamarck, J.-B. 1809. *Philosophie zoologique, ou exposition des considérations relatives à l'histoire naturelle des animaux* (Zoological philosophy: an exposition with regard to the natural history of animals). Trans. H. Elliot, Macmillan, London, 1914. Reprinted by University of Chicago Press, Chicago, IL, 1984.

Latham, A. 2005. *The Naked Emperor: Darwinism Exposed.* Janus Publishing Ltd, London.

Lawton, G. 2009. Why Darwin was wrong about the tree of life. *New Scientist* 2692: 34–39.

Lawton, J.H. and May, R.M. (eds.) 1995. *Extinction Rates.* Oxford University Press, Oxford.

Lederberg, J. and McCray, A.T. 2001. 'Ome sweet 'omics – a genealogical treasury of words. *The Scientist* 15: 8–9.

Levinton, J. 1994. Punctuated debate. *Nature* 368: 407.

Litchfield, H.E. (ed.) 1915. *Emma Darwin: A Century of Family Letters, 1792–1896.* 2 vols. John Murray, London.

Llewelyn, J. 2008. *Margins of Religion: Between Kierkegaard and Derrida.* Indiana University Press, Bloomington, IN.

Lomborg, B. 2001. *The Skeptical Environmentalist.* Cambridge University Press, Cambridge, UK.

Lomborg, B. (ed.) 2004. *Global Crises, Global Solutions.* Cambridge University Press, Cambridge, UK.

Lyell, C. 1830–33. *Principles of Geology.* Penguin Books Ltd (Penguin Classics), London, 1997.

Lyell, C. 1832. *Principles of Geology, being an attempt to explain the former changes of the Earth's surface, by reference to causes now in operation. Vol. 2.* John Murray, London.

Mallet, J. 2008. Wallace and the species concept of the early Darwinians. In *Natural Selection and Beyond: The Intellectual Legacy of Alfred Russel Wallace* (Smith, C.R. and Beccaloni, G.W., eds.). Oxford University Press, Oxford, pp. 102–113.

Malthus, T. 1798, 1830. *An Essay on the Principle of Population*. Penguin Books Ltd, London, 1985.

Marques, A.C., Dupanloup, I., Vinckenbosch, N., Reymond, A. and Kaessmann, H. 2005. Emergence of young human genes after a burst of retroposition in primates. *PLoS Biology* 3(11): e357 doi:10.1371/journal.pbio.0030357.

Mascaro, J. (trans.) 1962. *The Bhagavad Gita*. Penguin Books Ltd, London.

Mayden, R.L. 1997. A hierarchy of species concepts: the denouement in the saga of the species problem. In *Species: The Units of Diversity* (Claridge, M.F., Dawah, H.A. and Wilson, M.R., eds.). Chapman & Hall, London, pp. 381–384.

Mayr, E. 1942. *Systematics and the Origin of Species*. Columbia University Press, New York.

Mayr, E. 1991. *One Long Argument*. Harvard University Press, Cambridge, MA.

Mayr, E. 2004. *What Makes Biology Unique?* Cambridge University Press, Cambridge, UK.

McCafferty, D.J. 2009. Darwin in Scotland. *The Glasgow Naturalist* 25: 1–3.

McGrath, A. 2005. *Dawkins' God: Genes, Memes, and the Meaning of Life*. Blackwell Publishing, Oxford.

McGrath, A. 2007. *The Dawkins Delusion*. Society for Promoting Christian Knowledge, London.

Medawar, P. 1984. *The Limits of Science*. Oxford University Press, Oxford.

Mendel, G. 1865. Versuche über Pflanzen-Hybriden (Experiments with plant hybrids). *Transactions of the Brünn Natural History Society*.

Miller, G.F. 2000. Evolution of human music through sexual selection. In *The Origins of Music* (Wallin, N.L., Merker, B, and Brown, S., eds.). MIT Press, Cambridge, MA, pp. 329–360.

Miller, J. and Van Loon, B. 1982. *Darwin for Beginners*. Writers and Readers Publishing Cooperative Society, London.

Milne, D. 1847. On the parallel roads of Lochaber, with remarks on the change on relative levels of sea and land in Scotland, and on the detrital deposits of that county. *Transactions of the Royal Society of Edinburgh* 16: 395–418.

Musical Times 1901. *The Musical Times and Singing Class Circular*, Vol. 42, No. 700, June 1, pp. 369–374. Musical Times Publications Ltd, Berkhamsted, Hertfordshire.

Nagylaki, T. 1989. Gustave Malécot and the transition from classical to modern population genetics. *Genetics* 122: 253–268.

Nesse, R.M. and Williams, G.C. 1994. *Why We Get Sick: The New Science of Darwinian Medicine*. Times Books, Random House, New York.

New Lanark Conversation Trust 1997. *The Story of Robert Owen*. New Lanark Conversation Trust, New Lanark.

Nordborg, M. 2001. Introduction to coalescent theory. In *Handbook of Statistical Genetics* (Balding, D., Bishop, M. and Cannings, C., eds.). John Wiley & Sons, Chichester, pp. 179–212.

Nosil, P. and Sandoval, C.P. 2008. Ecological niche dimensionality and the evolutionary diversification of stick insects. *PLoS ONE* 3: e1907. doi:10.1371/

journal.pone.0001907.

Occhiogrosso, P. 1996. *The Joy of Sects*. Image Books, New York.

O'Hanlon, R. 1984. *Joseph Conrad and Charles Darwin: The Influence of Scientific Thought on Conrad's Fiction*. Salamander Press, Edinburgh.

Okasha, S. 2002. *Philosophy of Science*. Oxford University Press, Oxford.

Owen, R. 1853. Osteological contributions to the natural history of the chimpanzees and orangs. *Transcripts of the Zoological Society* 4: 75–88.

Owen, R. 1855. On the anthropoid apes and their relations to man. *Proceedings of the Royal Institute* 2: 26–41.

Owen, R. 1858. On the characters, principles of division and primary groups of the class Mammalia. *Journal of the Linnean Society* 2: 1–37.

Owen, R. 1860. Review of Origin & other works. *Edinburgh Review* 111: 487–532.

Owen, R. 1861. The gorilla and the negro. *Athenaeum* (23 March): 395–396.

Owen, R. 1863. On the *Archaeopteryx* of von Meyer, with a description of the fossil remains of a long-tailed species, from the Lithographic Slate of Solenhofen. *Philosophical Transactions of the Royal Society of London* 153: 33–47.

Owen, R. 1865. *Memoir on the Gorilla (Troglodytes Gorilla, Savage)*. Taylor and Francis, London.

Paley, W. 1802. *Natural Theology*. Oxford University Press, Oxford, 2006.

Palmer, D. 2005. *Seven Million Years: The Story of Human Evolution*. Weidenfeld & Nicolson, London.

Patterson, C. 1978. *Evolution*. British Museum (Natural History), London.

Pearson, P.N. 1996. Charles Darwin on the origin and diversity of igneous rocks. *Earth Sciences History* 15: 49–67.

Pearson, P.N. 2003. In retrospect: An Investigation into the Principles of Knowledge and of the Progress of Reason, from Sense to Science and Philosophy, by James Hutton 1794. *Nature* 425: 665.

Pearson, P.N. and Nicholas, C.J. 2007. 'Marks of Extreme Violence': Charles Darwin's geological observations on St Jago (São Tiago), Cape Verde Islands. In *Four Centuries of Geological Travel: The Search for Knowledge on Foot, Bicycle, Sledge and Camel* (Wyse-Jackson, P., ed.). Special Publication 287. Geological Society, London, pp. 239–253.

Peirce, C.S. 1965. How to make our ideas clear. In *Collected Papers: Vol. 5: Pragmatism and Pragmaticism, and Vol. 6: Scientific Metaphysics (two volumes in one)* (Hartshorne, C. and Weiss, P., eds.). The Belknap Press and Harvard University Press, Cambridge, MA, p. 268.

Petto, A.J. and Godfrey, L.R. (eds.) 2007. *Scientists Confront Creationism: Intelligent Design and Beyond*. W.W. Norton & Co., New York.

Polkinghorne, J. 1998. *Belief in God in an Age of Science*. Yale University Press, Yale, CT.

Pomiankowski, A. and Iwasa, Y. 1993. Evolution of multiple sexual preferences by Fisher's runaway process of sexual selection. *Proceedings of the Royal Society of London B* 253: 173–181.

Popper, K. 1961. *The Poverty of Historicism*. Routledge and Kegan Paul, London.

Quadri, M.M. 1967. *Integration of Islam with Science*. Ripon Printing Press Ltd, Ripon, WI.

Quammen, D. 2004. Was Darwin wrong? *National Geographic* 206: 2–35.

Raby, P. 2002. *Alfred Russel Wallace: A Life*. Pimlico, London.

Rogers, J.A. 1974. Russian opposition to Darwinism in the nineteenth century. *Isis* 65: 487–505.

Ronshaugen, M., McGinnis, N. and McGinnis, W. 2002. Hox protein mutation and macroevolution of the insect body plan. *Nature* 415: 914–917.

Rosner, L. 1992. Thistle on the Delaware: Edinburgh medical education and Philadelphia practice, 1800–1825. *Social History of Medicine* 5: 19–42.

Sardar, Z. and Wyn Davies, M. 2004. *Islam*. New Internationalist Publications Ltd, Oxford.

Schrödinger, E. 1944. *What Is Life?* Cambridge University Press, Cambridge, UK, 1992.

Sclater, A. 2006. The extent of Charles Darwin's knowledge of Mendel. *Journal of Bioscience* 31: 191–193.

Secord, J.A. 1991. Edinburgh Lamarckians: Robert Jameson and Robert E. Grant. *Journal of the History of Biology* 24: 1–18.

Secord, J.A. 2000. *Victorian Sensation: The Extraordinary Publication, Reception, and Secret Authorship of Vestiges of the Natural History of Creation*. University of Chicago Press, Chicago, IL.

Sexton, E. 2001. *Dawkins and the Selfish Gene*. Icon Books Ltd, Cambridge, UK.

Shermer, M. 2002. *In Darwin's Shadow: The Life and Science of Alfred Russel Wallace*. Oxford University Press, Oxford.

Shiga, D. 2007. Tight-knit trio of quasars discovered. *New Scientist*, http://www.newscientist.com/article/dn10915.

Skinner, B.F. 1957. *Verbal Behavior*. Copley Publishing Group, Acton, MA.

Smith, J., Jr. 1830. *Book of Mormon: An Account Written by the Hand of Mormon, Upon Plates Taken from the Plates of Nephi*. E.B. Grandin, Palmyra, NY.

Smith, P. 1996. *A Short History of the Bahá'í Faith*. One World Publications, Oxford.

Spadafora, C. 2008. Sperm-mediated 'reverse' gene transfer: a role of reverse transcriptase in the generation of new genetic information. *Human Reproduction* 23: 735–740.

Stangroom, J. 2005. *What Scientists Think*. Routledge, London.

Stark, J. 1806. *Picture of Edinburgh; Containing a History and Description of the City, with a Particular Account of Every Remarkable Object in, or Establishment Connected with, the Scottish Metropolis*. John Murray, London.

Steele, E.J. 2008. Reflections on the state of play in somatic hypermutation. *Molecular Immunology* 45: 2723–2726.

Sterelny, K. 2001. *Dawkins vs. Gould: Survival of the Fittest*. Icon Books Ltd, Cambridge, UK.

Stevenson, R.L. 1907. *Memoir of Fleeming Jenkin: Records of a Family of Engineers*. Scribner, New York.

Stewart, I. 2003. Self-organization in evolution: a mathematical perspective. *Philosophical Transactions of the Royal Society A* 361: 1101–1123.

Stewart, I., Elmhirst, T. and Cohen, J. 2003. Symmetry-breaking as an origin of species. In *Bifurcations, Symmetry, and Patterns* (Buescu, J., Castro, S. and Dias, A.P.S., eds.). Birkhauser, Basel, pp. 3–54.

Stott, R. 2003. *Darwin and the Barnacle*. Faber & Faber Ltd, London.

Swift, D.W. 2002. *Evolution Under the Microscope*. Leighton Academic Press, Stirling.

Thompson, H. 2005. *This Thing of Darkness*. Headline Review, London.

Thompson, M. 1997. *Philosophy of Religion*. Hodder Headline, London.

Thomson, K. 2005. *Fossils*. Oxford University Press, Oxford.

Tort, P. (ed.) 1996. *Dictionnaire du Darwinisme et de l'évolution*. 3 vols. PUF, Paris.

Tort, P. 2001. *Charles Darwin: The Scholar Who Changed Human History*. Thames & Hudson, London.

Tort, P. 2002. *La seconde révolution Darwinienne*. Kimé, Paris.

Tort, P. 2004. *Darwin et la philosophie*. Kimé, Paris.

Tyson, E. 1699. *Orang-outang, sive, Homo sylvestris: or, The anatomy of a pygmie compared with that of a monkey, an ape, and a man*. Printed for Thomas Bennet and Daniel Brown, London.

von Humboldt, A. 1814–25. *Personal Narrative of a Journey to the Equinoctial Regions of the New Continent*. Penguin Books Ltd (Penguin Classics), London, 1995.

Waddington, C.H. 1958. Theories of evolution. In *A Century of Darwin* (Barnett, S.A., ed.). Heinemann, London, p. 5.

Wallace, D.R. 1983. *The Klamath Knot: Explorations of Myth and Evolution*. Sierra Club Books, San Francisco, CA.

Watch Tower Bible and Tract Society of Pennsylvania 2000. *Jehovah's Witnesses: Who Are They? What Do They Believe?* Watch Tower Bible and Tract Society of New York, Inc., Brooklyn, NY.

Watterson, G.A. 1996. Motoo Kimura's use of diffusion theory in population genetics. *Theoretical Population Biology* 49: 154–188.

Watson, J.D. 2000. *A Passion for DNA: Genes, Genomes, and Society*. Oxford University Press, Oxford.

White, G. 1789. *The Natural History and Antiquities of Selborne*. Cassell & Company, London.

Wiley 2006. *Encyclopedia of Life Sciences*. John Wiley & Sons, Chichester.

Wilkins, J.S. 2002. *Summary of 26 Species Concepts*. http://www.utm.utoronto.ca/~w3bio/bio443/seminar_papers/summary_of_26_species_concepts.pdf

Wilkinson, D. 2002. Ecology before ecology: biogeography and ecology in Lyell's 'Principles'. *Journal of Biogeography* 29: 1109–1115.

Williams, G.C. and Nesse, R.M. 1991. The dawn of Darwinian medicine. *Quarterly Review of Biology* 66: 1–22.

Williams, R. 2006. *Unintelligent Design: Why God Isn't As Smart As She Thinks She Is*. Allen & Unwin, London.

Willis, J.C. 1940. *The Course of Evolution by Differentiation or Divergent Mutation rather than by Selection*. Cambridge University Press, Cambridge, UK.

Wilson, E.O. 1992. *The Diversity of Life*. Harvard University Press, Cambridge, MA.

Wilson, E.O. 1998. *Consilience: The Unity of Knowledge*. Knopf, New York.

Wilson, E.O. (ed.) 2006. *From So Simple a Beginning: The Four Great Books of Charles Darwin*. W.W. Norton & Co., New York.

Wolfram, S. 2002. *A New Kind of Science*. Wolfram Media Inc., Champaign, IL.

Wolpert, L. 2006. *Six Impossible Things Before Breakfast: The Evolutionary Origins of Belief*. Faber & Faber Ltd, London.

Wood, B. 2005. *Human Evolution*. Oxford University Press, Oxford.

Wool, D. 2001. Charles Lyell – "the father of geology" – as a forerunner of modern ecology. *Oikos* 94: 385–391.

World Heritage Steering Group 2006. *World Heritage Nomination: Darwin at Downe*. Department for Culture, Media and Sport, London.

Wyn Davies, M. 2000. *Darwin and Fundamentalism*. Icon Books Ltd, Cambridge, UK.

Zaehner, R.C. 1962. *Hinduism*. Oxford University Press, Oxford.

Zirkle, C. 1941. Natural selection before the "Origin of Species". *Proceedings of the American Philosophical Society* 84: 71–123.

Contributor Biographies

Nick Barton joined the University of Edinburgh in 1990, and has been a Professor of Evolutionary Biology since 1994 in which year he was also made a Fellow of the Royal Society; a year later he was made a Fellow of the Royal Society of Edinburgh. He received a Wolfson Merit Award in 2005, the Royal Society Darwin Medal in 2006 and the Darwin-Wallace Award in 2008. His research interests centre on understanding the evolution of traits which depend on interactions between large numbers of genes, and he is especially renowned for his work on hybrid zones often using toads, butterflies and grasshoppers: narrow regions where different populations meet and hybridise, used to find the genetic basis of species differences, and to measure rates of selection and gene flow. He recently published a definitive textbook *Evolution* (2007, with Derek E.G. Briggs, Jonathan A. Eisen, David B. Goldstein and Nipam H. Patel), and is now based at the Institute of Science and Technology, Austria.

Michael J. Behe is an American biochemist and Intelligent Design advocate. He currently serves as Professor of Biochemistry at Lehigh University in Pennsylvania and as a senior fellow of the Discovery Institute's Center for Science and Culture. Behe is best known for his argument for Irreducible Complexity, a concept that asserts that some structures are too complex at the biochemical level to be adequately explained as a result of evolutionary mechanisms and thus are the result of Intelligent Design. He is the author of *Darwin's Black Box* (1996).

Stuart Blackman received his doctorate from the University of Edinburgh in 1995. He now shares a 'particular interest in the relationship between science and politics' with Ben Pile, his co-editor of *Climate Resistance: Challenging Climate Orthodoxy* (http://www. climate-resistance.org/) which they base on the argument that 'Environmentalism is in the ascendant. It holds that instead of buffering ourselves against whatever Mother Nature has to throw at us, we should try to make the weather marginally different by cutting down on the things that make life worth living'.

Mark Blaxter started at the University of Edinburgh in 1978, and returned in 1995. He has been a Professor of Evolutionary Genomics since 2004. Since turning from evolutionary zoology to genomics biology, his major interests have lain in the genomics of 'neglected' animal phyla. His lab group uses modern sequencing technologies to generate expressed sequence tag and genome sequence data for non-vertebrates (such as earthworms, nematodes and tardigrades), and analyses these with a variety of bioinformatics tools, about which they run the website *Nematode and Neglected Genomics* (http://www.nematodes.org/).

Brian Charlesworth joined the University of Edinburgh in 1997 as a Professor of Evolutionary Biology. He became Head of Institute in 2007. His awards include: Fellow of the Royal Society (1991), Honorary Fellow of the American Academy of Arts and Sciences (1996), President, Society for the Study of Evolution (1999), Fellow of the Royal Society of Edinburgh (2000), Darwin Medal of the Royal Society (2000), President, UK Genetics Society (2006), Sewall Wright Award, American Society of Naturalists (2006), and the Frink Award, Zoological Society of London (2007). He has written *Evolution: A Very Short Introduction* (2003, with Deborah Charlesworth) and *Evolution in Age-structured Populations* (2008, with C. Cannings, Frank C. Hoppensteadt and L.A. Segel).

Noam Chomsky presented the University of Edinburgh Gifford Lecture 2004/05. He is an Institute Professor Emeritus and Professor Emeritus of Linguistics at the Massachusetts Institute of Technology. Since he began his theory of generative grammar, his works have had a major influence on theoretical linguistics and cognitive science. He is a prominent critic of US foreign and domestic policy, and has been cited more often than any other living scholar between 1980 and 1992, and was ranked the eighth most-cited ever. He has written about 30 books on linguistics, one on computer science and over 70 political works, most recently, in each category, *New Horizons in the Study of Language and Mind* (2000), *Three Models for the Description of Language* (1956) and *What We Say Goes: Conversations on U.S. Power in a Changing World* (2007), respectively.

Richard Dawkins is a regular speaker at the Edinburgh Book Festival and has taken part in the Edinburgh International Science Festival. He held the Charles Simonyi Chair of Public Understanding of Science since its inauguration in 1995 until his retirement in 2008. His many awards include the Royal Society of Literature Award (1987), Los Angeles Times Literary Prize (1987), Sci.Tech Prize for Best Television Documentary Science Programme of the Year (1987), Zoological Society of London Silver Medal (1989), Royal Society of London Michael Faraday Award (1990), Nakayama Prize for Achievement in Human Science (1994), Humanist of the Year Award (1996), International Cosmos Prize, Osaka, Japan (1997), Kistler Prize, USA (2001), Medal of the Presidency of the Italian Republic, Rimini, Italy (2001), Bicentennial Kelvin Medal, Royal Philosophical Society of Glasgow (2002), and the Shakespeare Prize for Contribution to British Culture (2005). He presented the Royal Institution Christmas Lecture 'Growing Up in the Universe' in 1991 and was elected a Fellow of the Royal Society of Literature in 1997 and a Fellow of the Royal Society in 2001. He frequently appears in the media, most recently in his TV series *The Genius of Charles Darwin*

(2008). His first book was famously *The Selfish Gene* (1976), and his most notorious *The God Delusion* (2006). His most recent is *The Greatest Show on Earth* (2009).

William A. Dembski is a mathematician and philosopher and Research Professor in Philosophy at Southwestern Baptist Theological Seminary in Fort Worth. He is also a senior fellow with the Discovery Institute's Center for Science and Culture in Seattle. Previously he was the Carl F.H. Henry Professor of Theology and Science at The Southern Baptist Theological Seminary in Louisville, where he founded its Center for Theology and Science. Before that he was Associate Research Professor in the Conceptual Foundations of Science at Baylor University, where he also headed the first Intelligent Design think-tank at a major research university: The Michael Polanyi Center. His books include *The Design Revolution* (2004), *Understanding Intelligent Design* (2008, with Sean McDowell) and *The End of Christianity* (2009).

Daniel C. Dennett presented the University of Edinburgh Nature of Knowledge Lecture in 2006 and returned to give the Enlightenment Lecture in 2007 when he was also awarded with an Honorary Degree of Doctor of Letters. His many other awards include the A.P.A. Barwise Prize (2004), Humanist of the Year from the American Humanist Association (2004), Bertrand Russell Society Award (2004), Academy of Achievement Golden Plate Award (2006) and the Richard Dawkins Prize from the Atheist Alliance International (2007). He has been the co-director of the Center for Cognitive Studies and the Austin B. Fletcher Professor of Philosophy at Tufts University since 2000, and became the Leverhulme Professor in the Department of Philosophy and History of Science at the London School of Economics in 2001. Of his many books, his most recent is *Breaking the Spell: Religion as a Natural Phenomenon* (2006).

Alistair Gentry was artist in residence for New Media Scotland and the ESRC Genomics Policy and Research Forum at the University of Edinburgh (2006–2007). He is a 'free range' writer and artist producing work for publication, performance, broadcast and installation, as digital media, on radio, television and the stage, in art galleries, at film festivals, in print and on the Net, and has been described as 'subversive' (*SFX*), 'startling' (*The Independent*) and 'fascinating' (*The Times*). His books include *Three Times True* (2007) featuring his work at the university, and his latest *Uncanny Valley* (2008).

Richard L. Gregory co-founded the Department of Machine Intelligence and Perception, a forerunner of the Department of Artificial Intelligence at the University of Edinburgh (with Donald Michie and Christopher Longuet-Higgins), in 1967, the same year that he presented the Royal Institution Christmas Lecture, 'The Intelligent Eye'. He was made a Fellow of the Royal Society of Edinburgh in 1969, and a CBE in 1989, and elected to be a Fellow of the Royal Society in 1992, the same year that he was awarded the Royal Society Michael Faraday Medal, while he received the Hughling Jackson Gold Medal from the Royal Society of Medicine in 1999. He is currently an Emeritus Professor of Neuropsychology at the University of Bristol. His contributions to TV and radio are extensive, as are his contributions to the design of science exhibitions, most notably *Hands-On Science* at the Bristol 'Exploratory'. His books include *Mirrors in Mind* (1997) and *The Mind Makers* (1998).

Ken Ham toured Scotland in 2007, as part of which he delivered a sermon on Young Earth Creationism and the literal interpretation of the Book of Genesis in Edinburgh, as the President of Answers in Genesis-U.S. and Joint CEO of Answers in Genesis International, and has been instrumental in setting up the Creation Museum in Petersburg, Kentucky. He frequently appears in the media and presents the daily radio and internet programme *Answers ... with Ken Ham*. His books include *Refuting Evolution* (1999, with Jonathan Sarfati), *The New Answers Book* (2006) and the latest *Dinosaurs for Kid*s (2009).

Adrian Hawkes helped to form Phoenix Community Care Ltd (which looks after some 30+ unaccompanied minors, and vulnerable adults, in housing in North London), the PCC Foster Care agency and the London Training Consortium Ltd which trains refugees and asylum seekers in ESOL, IT, and literacy. His books include *Jacob – A Fatherless Generation* (2002), *Hello!: Is That You God?* (2007) and his latest, *Culture Clash* (2009).

Richard Holloway is a writer and broadcaster, and was Bishop of Edinburgh from 1986, in the Scottish Episcopal Church, of which he was elected Primus in 1992, but resigned in 2000. He was elected Fellow of the Royal Society of Edinburgh in 1995, and was Professor of Divinity at Gresham College in London (1997–2001), member of the Human Fertilisation and Embryology Authority (1990–1997), and chair of the BMA Steering Group on Ethics and Genetics (1995–1998), and is currently a member of the Broadcasting Standards Commission, and chair of the Scottish Arts Council (since 2005) and of Sistema Scotland (since 2006). His latest book is *Between the Monster and the Saint* (2008).

Randal Keynes is a British conservationist and author and a great-great-grandson of Charles Darwin. He has been awarded the Order of the British Empire (OBE), is a central figure in the bid for Down House to be a World Heritage Site, has collaborated with *Darwin Online*, is a supporter of the Humanist Society, and is the acclaimed author of *Annie's Box: Darwin, His Daughter, and Human Evolution* (2001), upon which the film *Creation* (2009) was based.

Simon Kirby is Professor of the Evolution of Language and Cognition at the University of Edinburgh. He has written many academic articles and invited chapters, plus the well-received books, *Function, Selection and Innateness: The Emergence of Language Universals* (1999), *Language Evolution* (2003, with Morten Christiansen) and The *Evolution of Language* (2007).

Antony Latham is a doctor working in general practice in the Scottish Outer Hebrides. After qualifying in medicine at Trinity College Dublin, he worked for nine years in East Africa before settling with his family on the Isle of Harris. He is the author of *The Naked Emperor: Darwinism Exposed* (2005).

John Llewelyn has been Reader in Philosophy at the University of Edinburgh, Visiting Professor at the University of Memphis and Arthur J. Schmitt Distinguished Visiting Professor of Philosophy at Loyola University of Chicago. Among his publications are *Beyond Metaphysics?: The Hermeneutic Circle in Contemporary Continental Philosophy* (1985), *Derrida on the Threshold of Sense* (1986), *The Middle Voice of Ecological Conscience* (1991), *Emmanuel Levinas: The Genealogy of Ethics* (1995), *The HypoCritical Imagination* (2000), *Appositions of Jacques Derrida and Emmanuel Levinas* (2002), *Seeing Through God: A Geophenomenology* (2003), and *Margins of Religion: Between Kierkegaard and Derrida* (2008).

Aubrey Manning was Professor of Natural History at the University of Edinburgh from 1973 to 1997. He was elected Fellow of the Royal Society of Edinburgh in 1973, and received an OBE in 1998, and the Zoological Society of London Silver Medal in 2003 for public understanding of science. He has been Chairman of Edinburgh Brook Advisory Centre, Chairman of the Council of the Scottish Wildlife Trust and a trustee of the National Museums of Scotland and Project Wallacea. He is President of the Royal Society of Wildlife Trusts and Patron of the Optimum Population Trust. He has presented numerous TV and radio programmes on natural history, most recently *Earth Story*, *Talking Landscapes*, *The Sounds of Life* and *The Rules of Life*. He co-authored the classic text *An Introduction to Animal Behaviour* (1967, with Marian Stamp Dawkins).

George E. Marshall is a Professor of Molecular and Cellular Biology and an Honorary Lecturer at the Faculty of Biomedical and Life Sciences of the University of Glasgow. He has been the Sir Jules Thorn Lecturer in Ophthalmic Science. He came to prominence after providing an interview (*An Eye for Creation*) to the Answers in Genesis magazine *Creation* in 1996, refuting the idea that the human eye is 'wired backwards'.

Geoff Morgan is the Assistant Principal Teacher of Biology at George Watson's College, a leading co-educational independent day school in Edinburgh, as well as being an examiner for the Scottish Qualifications Authority (SQA) and an accomplished ornithologist, and exhibited amateur wildlife photographer.

Martyn Murray is a zoologist and lecturer at the University of Edinburgh as well as CEO of MGM Environmental Solutions Ltd, a leading provider of consultancy services in biodiversity conservation, founded in 1991. He was a Senior Research Fellow at the University of Cambridge, leading long-term projects in Tanzania on the community ecology and migrations of large mammals in the savanna and open grasslands of Serengeti National Park, and in Malaysia on the ecology of figs and figwasps. He also worked with the Global Security Programme to explore the consequences of environmental insecurity and free trade on the future of biodiversity, and he was Senior Advisor to the World Conservation and Monitoring Centre, and a member of the Species Survival Commission and the European Sustainable Use Specialist Group. He is the author of *The Storm Leopard* (Whittles Publishing, 2010).

Paul Pearson is Professor in Paleoclimatology and Director of Research in the School of Earth, Ocean and Planetary Sciences at Cardiff University. He is interested in micropalaeontology, palaeoceanography, geochemistry, evolutionary patterns and processes, often extracting climatic information from deep sea cores and sediments, and the history of science and archaeology. He has sailed on several occasions with the Ocean Drilling Program, but came to prominence when he wrote about James Hutton in *Nature*.

Martin Rees delivered the 2007 Gifford Lecture, '21st Century Science: Cosmic Perspectives and Terrestrial Challenges', at the University of St Andrews, and has taken part in the Edinburgh International Science Festival. His full form of address is 'Professor Sir Martin John Rees, Baron Rees of Ludlow, of Ludlow in the County of Shropshire'. He has been President of the Royal Society since 2005, and is also Master of Trinity College, and Professor of Cosmology and Astrophysics at the University of Cambridge, and Visiting Professor at Leicester University and Imperial College London. He was knighted in 1992, appointed Astronomer Royal in 1995, and was nominated to the House of Lords in 2005 as a cross-bench peer, and appointed a member of the Order of Merit in 2007. His current research deals with cosmology and astrophysics, especially gamma ray bursts, galactic nuclei, black hole formation and radiative processes (including gravitational waves) and also cosmic structure formation, especially the early generation of stars and galaxies that formed at the end of the cosmic dark ages more than 12 billion years ago relatively shortly after the Big Bang. His recent awards include the Royal Society's Michael Faraday Prize and lecture for science communication (2004), and the Royal Swedish Academy's Crafoord Prize (2005). Other notable awards include the Heinemann Prize (1984), the Balzan Prize (1989), the Bower Award of the Franklin Institute (1998), the Einstein Award from the World Cultural Council (2003) and the UNESCO Niels Bohr Medal (2005). He has authored or co-authored about 500 research papers. He has lectured, broadcast and written widely on science and policy, and is the author of seven books for a general readership, most recently *What We Still Don't Know* (2007).

Contributor Index

Subject Index

JF Derry

JF Derry is a writer who was previously a conservation ecologist and rangeland scientist mainly working in British and African universities, but also in Spain, Brussels, Mongolia and Australia. He had intended to discover the secrets to life with degrees in Biochemistry, Bioelectronics and Biological Computation, but somehow also ended up with a PhD in African Ecology, specialising in the consequences of savanna waterholes on plants and animals. Thus his publication history is mainly in academic journals, on aspects of computational biology, pastoralism and grazing studies, ergo a recent manuscript on evolution in African bovids (antelopes and cattle). He has written for several national newspapers, countless technical reports for various governments, and as a reviewer for several major record labels, the TLS and the university presses of Oxford and Cambridge and Harvard, several major and independent book publishers, and the National Portrait Gallery. He has also written extensively on Charles Darwin, and his recent poetry can be found in such places as the Human Genre Project. He once collected lion semen, intentionally. "How do you …?", you ask. The answer is, of course, "very carefully". JF lives in Edinburgh, with his partner and their two daughters.